NATIONAL ACADEMIES *Sciences*
Engineering
Medicine

NATIONAL ACADEMIES PRESS
Washington, DC

Protecting U.S. Technological Advantage

I0357551

Committee on Protecting Critical Technologies for National Security in an Era of Openness and Competition

Board on Science, Technology, and Economic Policy

Committee on Science, Technology, and Law

Committee on Science, Engineering, Medicine, and Public Policy

Policy and Global Affairs

Intelligence Community Studies Board

Division on Engineering and Physical Sciences

Consensus Study Report

NATIONAL ACADEMIES PRESS 500 Fifth Street, NW, Washington, DC 20001

This activity was supported by contracts between the National Academy of Sciences and the Defense Advanced Research Projects Agency (W911NF-18-D-0002/W911NF18F0089) and the National Science Foundation (SMA-1946465). Any opinions, findings, conclusions, or recommendations expressed in this publication do not necessarily reflect the views of any organization or agency that provided support for the project.

International Standard Book Number-13: 978-0-309-69130-7
International Standard Book Number-10: 0-309-69130-3
Digital Object Identifier: https://doi.org/10.17226/26647
Library of Congress Control Number: 2022948221

This publication is available from the National Academies Press, 500 Fifth Street, NW, Keck 360, Washington, DC 20001; (800) 624-6242 or (202) 334-3313; http://www.nap.edu.

Copyright 2022 by the National Academy of Sciences. National Academies of Sciences, Engineering, and Medicine and National Academies Press and the graphical logos for each are all trademarks of the National Academy of Sciences. All rights reserved.

Printed in the United States of America.

Suggested citation: National Academies of Sciences, Engineering, and Medicine. 2022. *Protecting U.S. Technological Advantage*. Washington, DC: The National Academies Press. https://doi.org/10.17226/26647.

The **National Academy of Sciences** was established in 1863 by an Act of Congress, signed by President Lincoln, as a private, nongovernmental institution to advise the nation on issues related to science and technology. Members are elected by their peers for outstanding contributions to research. Dr. Marcia McNutt is president.

The **National Academy of Engineering** was established in 1964 under the charter of the National Academy of Sciences to bring the practices of engineering to advising the nation. Members are elected by their peers for extraordinary contributions to engineering. Dr. John L. Anderson is president.

The **National Academy of Medicine** (formerly the Institute of Medicine) was established in 1970 under the charter of the National Academy of Sciences to advise the nation on medical and health issues. Members are elected by their peers for distinguished contributions to medicine and health. Dr. Victor J. Dzau is president.

The three Academies work together as the **National Academies of Sciences, Engineering, and Medicine** to provide independent, objective analysis and advice to the nation and conduct other activities to solve complex problems and inform public policy decisions. The National Academies also encourage education and research, recognize outstanding contributions to knowledge, and increase public understanding in matters of science, engineering, and medicine.

Learn more about the National Academies of Sciences, Engineering, and Medicine at **www.nationalacademies.org**.

Consensus Study Reports published by the National Academies of Sciences, Engineering, and Medicine document the evidence-based consensus on the study's statement of task by an authoring committee of experts. Reports typically include findings, conclusions, and recommendations based on information gathered by the committee and the committee's deliberations. Each report has been subjected to a rigorous and independent peer-review process and it represents the position of the National Academies on the statement of task.

Proceedings published by the National Academies of Sciences, Engineering, and Medicine chronicle the presentations and discussions at a workshop, symposium, or other event convened by the National Academies. The statements and opinions contained in proceedings are those of the participants and are not endorsed by other participants, the planning committee, or the National Academies.

Rapid Expert Consultations published by the National Academies of Sciences, Engineering, and Medicine are authored by subject-matter experts on narrowly focused topics that can be supported by a body of evidence. The discussions contained in rapid expert consultations are considered those of the authors and do not contain policy recommendations. Rapid expert consultations are reviewed by the institution before release.

For information about other products and activities of the National Academies, please visit www.nationalacademies.org/about/whatwedo.

COMMITTEE ON PROTECTING CRITICAL TECHNOLOGIES FOR NATIONAL SECURITY IN AN ERA OF OPENNESS AND COMPETITION

PATRICK D. GALLAGHER, *Co-chair*, Chancellor, University of Pittsburgh
SUSAN M. GORDON, *Co-chair*, Former Principal Deputy Director of National Intelligence
ROBERT J. BIRGENEAU (NAS), Chancellor Emeritus and Arnold and Barbara Silverman Professor of Physics, Materials Science and Engineering, and Public Policy, University of California, Berkeley
ROBERT C. DYNES (NAS), President Emeritus, University of California; and Professor, Department of Physics, University of California, San Diego
DEBORAH FRINCKE, Associate Labs Director, National Security Programs, Sandia National Laboratories
GILBERT HERRERA (Member 2/16/2021–8/15/2021), Laboratory Fellow, Sandia National Laboratories*
LEROY E. HOOD (NAS/NAE/NAM), Senior Vice President and Chief Science Officer, Providence St. Joseph Health; and Chief Strategy Officer, Cofounder, and Professor, Institute for Systems Biology
MICHAEL J. IMPERIALE, Arthur F. Thurnau Professor of Microbiology and Immunology, University of Michigan
J. MICHAEL MCQUADE, Special Advisor to the President, Carnegie Mellon University
JUDITH A. MILLER, Independent Consultant
RICHARD M. MURRAY (NAE), Thomas E. and Doris Everhart Professor of Control and Dynamical Systems and Bioengineering, California Institute of Technology

*Resigned from the committee effective August 15, 2021.

STAFF

GAIL E. COHEN, Senior Director, Board on Science, Technology, and Economic Policy (STEP), *Study Director*
ANNE-MARIE MAZZA, Senior Director, Committee on Science, Technology, and Law, *Co–Study Director* (through June 2021)
MEGHAN ANGE-STARK, Program Officer, STEP Board (through July 2021)
SOPHIE BILLINGE, Senior Program Assistant, STEP Board (through June 2022)
DAVID DIERKSHEIDE, Program Officer, STEP Board
BIANCA ESPINOSA, Christine Mirzayan Science and Technology Fellow, Committee on Science, Engineering, Medicine, and Public Policy

BOARD ON SCIENCE, TECHNOLOGY, AND ECONOMIC POLICY

ADAM B. JAFFE, *Chair*, Brandeis University
NOËL BAKHTIAN, Lawrence Berkeley National Laboratory
JEFF BINGAMAN, Former U.S. Senator, New Mexico
BRENDA J. DIETRICH (NAE), Cornell University
BRIAN G. HUGHES, HBN Shoe, LLC, San Antonio, Texas
PAULA E. STEPHAN, Georgia State University
SCOTT STERN, Massachusetts Institute of Technology
JOHN C. WALL (NAE), Cummins, Inc. (Retired)
JOHN L. ANDERSON (NAE), *Ex Officio Member*, National Academy of Engineering
VICTOR J. DZAU (NAM), *Ex Officio Member*, National Academy of Medicine
MARCIA MCNUTT (NAS/NAE), *Ex Officio Member*, National Academy of Sciences

STAFF
GAIL COHEN, Senior Director
DAVID DIERKSHEIDE, Program Officer
CLARA SAVAGE, Financial Officer

COMMITTEE ON SCIENCE, TECHNOLOGY, AND LAW

DAVID BALTIMORE (NAS/NAM), *Co-chair*, California Institute of Technology
DAVID S. TATEL, *Co-chair*, U.S. Court of Appeals for the District of Columbia
ERWIN CHEMERINSKY, University of California, Berkeley
ELLEN W. CLAYTON (NAM), Vanderbilt University Medical Center
JOHN O. DABIRI, California Institute of Technology
JENNIFER EBERHARDT (NAS), Stanford University
FEI-FEI LI (NAE/NAM), Stanford University
JUDITH A. MILLER, Independent Consultant
MARTHA L. MINOW, Harvard Law School
KIMANI PAUL-EMILE, Fordham University
NATALIE RAM, University of Maryland Carey School of Law
LISA RANDALL (NAS), Harvard University
PAUL M. ROMER, New York University
WILLIAM B. SCHULTZ, Zuckerman Spaeder LLP
JOSHUA M. SHARFSTEIN (NAM), Johns Hopkins Bloomberg School of Public Health
SUSAN S. SILBEY, Massachusetts Institute of Technology
GREGORY STONE, Tolles & Olson, LLP
JOHN S. COOKE, *Ex Officio Member*, The Federal Judicial Center

STAFF
ANNE-MARIE MAZZA, Senior Director
STEVEN KENDALL, Program Officer
DOMINIC LOBUGLIO, Senior Program Assistant

COMMITTEE ON SCIENCE, ENGINEERING, MEDICINE, AND PUBLIC POLICY

ALAN I. LESHNER (NAM), American Association for the Advancement of Science (Retired)
CLAIRE D. BRINDIS (NAM), University of California, San Francisco
KATHARINE G. FRASE (NAE), International Business Machines Corporation
JOHN G. HILDEBRAND (NAS), University of Arizona
EDWARD D. LAZOWSKA (NAE), University of Washington
FRANCES S. LIGLER (NAE), Texas A&M Health Sciences Center
JUANITA L. MERCHANT (NAM), University of Arizona College of Medicine
RICHARD A. MESERVE (NAE), Carnegie Institution for Science
C. PAUL ROBINSON (NAE), Sandia National Laboratories (Retired)
ROBERT F. SPROULL (NAE), University of Massachusetts at Amherst
JAMES M. TIEN (NAE), University of Miami
RUTH J. WILLIAMS (NAS), University of California, San Diego
SUSAN M. WOLF (NAM), University of Minnesota, Minneapolis
JOHN L. ANDERSON (NAE), *Ex Officio Member*, National Academy of Engineering
VICTOR J. DZAU (NAM), *Ex Officio Member*, National Academy of Medicine
MARCIA MCNUTT (NAS/NAE), *Ex Officio Member*, National Academy of Sciences

STAFF
TOM WANG, Senior Board Director
TOM ARRISON, Senior Advisor
COLE DONOVAN, Senior Program Officer
SARAH ROVITO, Senior Program Officer

INTELLIGENCE COMMUNITY STUDIES BOARD

MARK M. LOWENTHAL, *Co-chair*, Intelligence & Security Academy, LLC
MICHAEL A. MARLETTA (NAS/NAM), *Co-chair*, University of California, Berkeley
JOEL BRENNER, Joel Brenner, LLC
ROBERT CARDILLO, The Cardillo Group, LLC
FREDERICK R. CHANG (NAE), Southern Methodist University
DEAN B. CHENG, The Heritage Foundation
ROBERT C. DYNES (NAS), University of California, San Diego
ROBERT A. FEIN, McLean Hospital and Harvard Medical School
HUBAN A. GOWADIA, Lawrence Livermore National Laboratory
MARGARET A. HAMBURG (NAM), Nuclear Threat Initiative
MIRIAM E. JOHN, Independent Consultant
ANITA K. JONES (NAE), University of Virginia (Emerita)
STEVEN E. KOONIN (NAS), New York University
CARMEN L. MIDDLETON, The Walt Disney Company
ARTHUR L. MONEY (NAE), Independent Consultant
WILLIAM C. OSTENDORFF, United States Naval Academy
DAVID A. RELMAN (NAM), Stanford University and VA Palo Alto Health Care System
ELIZABETH RINDSKOPF PARKER, State Bar of California
SAMUEL S. VISNER, MITRE Corporation and Georgetown University
DAVID A. WHELAN (NAE), Cubic

STAFF
CARYN LESLIE, Acting Director
DIONNA ALI, Associate Program Officer
BRYAN BUNNELL, Research Associate
TONY FAINBERG, Senior Program Officer
NIA JOHNSON, Program Officer
CHRIS JONES, Senior Financial Manager
MARGUERITE SCHNEIDER, Administrative Coordinator

Preface

U.S. leadership in technology innovation is central to our nation's interests, including its security, economic prosperity, and quality of life. Our nation has created a science and technology ecosystem that fosters innovation, risk taking, and the discovery of new ideas that lead to new technologies through robust collaborations across and within academia, industry, and government, and our research and development enterprise has attracted the best and brightest scientists, engineers, and entrepreneurs from around the world. The quality and openness of our research enterprise have been the basis of our global leadership in technological innovation, which has brought enormous advantages to our national interests.

The committee's task was to examine and evaluate the need for boundaries or protections on the openness of scientific research and take into account the benefits and drawbacks of technology protection options. Heightened concerns about potential loss of leadership in critically important technology areas have led to increased rhetoric about the need to escalate protection and restrictions for certain technologies, but in an increasingly competitive and technology-dependent world, ensuring and protecting the nation's ability to lead in technological innovation is of critical importance. Given changes in technology and the global, interconnected competitive environment, the committee found that protecting technologies themselves is often ineffective or even counterproductive.

Instead, the committee believes that a fundamental shift is needed—one that moves away from specific technology controls to a risk management approach that focuses on protecting U.S. advantages in technology leadership and development. Strategies are needed for maximizing our advantages, promoting the scale and speed of our research and technology innovation ecosystem, fostering a risk-taking environment, and attracting, retaining, and supporting the most talented science, engineering, and innovation workforce in the world.

The committee also recognizes that new technologies are increasingly being developed on shared platforms. These platforms speed the scope and scale of new technologies, but they also have unique vulnerabilities associated with them. The committee recommends the development of a new multisector, multiorganizational, multinational approach to both protection and assurance of these platforms.

Legislation passed after the committee finished its deliberations—namely the CHIPS and Science Act—contains provisions that the committee hopes will facilitate its recommendations: expanding the support of the National Institute of Standards and Technology for standards capacity building and recognizing the critical importance of STEM (science, technology, engineering, and mathematics) graduate students and the STEM workforce as a whole.

Collectively, our recommendations are directed at building a healthier, more effective, and more resilient research and development ecosystem.

ACKNOWLEDGMENTS

We are deeply indebted to the hard work of the committee, which reviewed papers; engaged in thoughtful deliberations with speakers from academia, industry, and government; and spent considerable time developing findings and recommendations. Invaluable help was provided by the consultant writer, Steve Olson. The report also benefited from the input of Evan Johnson and his associates, who aided with the preparation of figures in the report. We are grateful for the dedication of the National Academies staff: Gail Cohen, David Dierksheide, Anne-Marie Mazza, Meghan Ange-Stark, and Sophie Billinge. We acknowledge all with deep gratitude.

Patrick Gallagher   Susan M. Gordon

Acknowledgment of Reviewers

This Consensus Study Report was reviewed in draft form by individuals chosen for their diverse perspectives and technical expertise. The purpose of this independent review is to provide candid and critical comments that will assist the National Academies of Sciences, Engineering, and Medicine in making each published report as sound as possible and to ensure that it meets the institutional standards for quality, objectivity, evidence, and responsiveness to the study charge. The review comments and draft manuscript remain confidential to protect the integrity of the deliberative process.

We thank the following individuals for their review of this report: Arthur Bienenstock, Stanford University; Vinton Cerf, Google, LLC; Jennie Hwang, H-Technologies Group, Inc.; Eric Isaacs, Carnegie Institution for Science; Marc Kastner, Massachusetts Institute of Technology; Farrokh Khatibi, Qualcomm; Theodore Sizer, Nokia Bell Labs; Sridhar Tayur, Carnegie Mellon University; Mitchel Wallerstein, Baruch College of the City University of New York; and Michael Wertheimer, The Chertoff Group.

Although the reviewers listed above provided many constructive comments and suggestions, they were not asked to endorse the conclusions or recommendations of this report, nor did they see the final draft before its release. The review of this report was overseen by Eric Kaler, Case Western Reserve University, and Catherine Novelli, Georgetown University. They were responsible for making certain that an independent examination of this report was carried out in accordance with the standards of the National Academies and that all review comments were carefully considered. Responsibility for the final content rests entirely with the authoring committee and the National Academies.

Contents

SUMMARY		**1**
1	**INTRODUCTION**	**11**
	Context for This Study, 11	
	Study Purpose, Charge, and Approach, 13	
	New Policies for a New Era, 15	
	Structure of the Report, 16	
2	**CHANGES IN TECHNOLOGY DEVELOPMENT AND COMMERCIALIZATION**	**19**
	How Technology Development and Commercialization Have Changed, 21	
	Case Study: Microelectronics, 27	
	Case Study: Artificial Intelligence, 32	
	Case Study: Synthetic Biology, 36	
	Case Study: Quantum Computing and Quantum Information Science, 41	
	Implications for Policy and Practice, 44	
3	**THE NEW COMPETITIVE LANDSCAPE**	**45**
	The Competitive Environment in the 1950–1985 Timeframe, 45	
	The Cold War National Security Competition, 48	
	The Resulting Policy Landscape, 49	
	Today's Competitive Landscape, 55	
	The Expansion of Control Mechanisms, 59	
	Implications of the New Competitive Landscape for U.S. Policies and Procedures, 64	
4	**THE COMPETITIVE CHALLENGE POSED BY CHINA**	**67**
	Features of the Competition between the United States and China, 69	
	Synthetic Biology in China, 71	

China's Activities in Microelectronics, Artificial Intelligence, and Quantum Computing, 75
Human Resources in the United States and China, 77
Implications of China's Actions for the Protection of U.S. Interests, 82

5   FINDINGS   **83**

6   RECOMMENDATIONS   **91**
Maximization of Strengths in Science, Research, and Technology Innovation, 92
Developing and Attracting Talent, 95
Identification of Strategic Technologies and Coordinated Risk Management, 97
Tailored Approaches to the Unique Vulnerabilities Resulting from Shared Platforms, 99

REFERENCES   **103**

A   AGENDAS   **115**

B   BIOGRAPHIES OF COMMITTEE MEMBERS   **123**

Summary

U.S. leadership in technology innovation is central to the nation's interests, including its security, economic prosperity, and quality of life. The United States has enjoyed enormous benefits from its global technology leadership in the form of enhanced national security, economic growth, and a high standard of living and well-being for its citizens. Beginning in the aftermath of World War II, the United States built and maintained an innovation system comprising research, development, commercialization, and production of technology-based goods and services, and it has enjoyed a position of global dominance over its competitors. The clear benefits of this leadership for the nation, which many argue was a major factor in determining the outcome of the Cold War, have long attracted the attention of competitors and adversaries alike while motivating U.S. efforts to protect its technological advantage.

The approach to risk and protection of technology advantage taken by the United States has been predicated on an assumption of economic and technological dominance in many aspects of the nation's research and development (R&D) system. U.S. research laboratories and universities have been regarded as global leaders in advancing the scientific discoveries and technological breakthroughs that have led to the emergence of new and advanced technologies. Top science and engineering talent from around the world has considered the United States the default "go-to" destination because of the ability to innovate, collaborate, and discover in an open, welcoming, and first-rate research environment. The U.S. government or U.S.-based companies have often been the first to develop new technologies and deploy them in the market. In doing so, they have been able to shape market conditions, build the user base for new technologies, and create regulatory frameworks to support those technologies.

The U.S. approach to risk and protection of its technological advantage has also rested on an assumption that the set of technologies driving military competitiveness, such as those employed in spy satellites, remains relatively distinct from the set driving commercial products and markets, allowing the nation's strategically sensitive technologies to be protected through limits or controls on information about them and on their production or use. In other words, protecting U.S. *advantages* in technology has been primarily an exercise in protecting the technology *outputs* of the nation's innovation enterprise.

In today's rapidly changing landscapes of technology and competition, however, the assumption that the United States will continue to hold a dominant competitive position by depending primarily on its historical approach of identifying specific and narrow technology areas requiring controls or restrictions is not valid. Further challenging that approach is the proliferation of highly integrated and globally shared platforms that power and enable most modern technology applications. Accordingly, this report offers a number of recommendations designed to help ensure that in this new environment, the United States will continue to enjoy its fundamental advantages in technological leadership.

THE CHANGING NATURE OF TECHNOLOGY DEVELOPMENT AND APPLICATION AND THE RISE OF PLATFORMS

Science, technology, and innovation are much more multidisciplinary, interdependent, and multinational today than in the past, a shift that complicates efforts to protect individual technologies from competitors in either the military or commercial realm. Technology products used to be largely discrete, with well-defined purposes. In contrast, many of today's technologies are multipurpose; have diffuse origins; and are highly dependent on other technologies with owners, users, and stakeholders from multiple countries. As a result, the R&D process that creates new technologies has become much more collaborative and internationally distributed.

Over the past few decades, moreover, military technologies have become increasingly dependent on technology development conducted in the commercial sector. In many strategically important technology fields, such as artificial intelligence, synthetic biology, and microelectronics, the pathway from basic research to application starts with private-sector investments aimed at addressing commercial markets. Thus while in the past, technology tended to emerge from the military to find commercial application, commercial R&D has now become the driver of much military technology.

Finally, many new technologies are developed, commercialized, and produced using systems of enabling technologies. These "platforms"—sets of integrated technologies, with associated institutional and human infrastructure, that serve as an essential foundation for the design, development, production, or use of specific technology applications—are typically multiuse, multipurpose, and multinational systems with many potential applications, often at a global scale. They can be rapidly scaled and stacked upon each other or interconnected, thereby multiplying their effects. Examples of such platforms include operating systems, telecommunications networks (such as 5G), the internet, genome editing, and microelectronics fabrication technologies. Platforms enable rapid, massive-scale, and lower-cost development by incorporating shareable technology elements into new technology applications. They are often developed and operated by the private sector and have become an essential part of the technology ecosystem.

As suggested above, the growth of systems-based technologies and platforms is disrupting traditional approaches to technology protection. Because they are shared, such platforms cannot be protected using the historical approach of restricting knowledge or use without causing widespread problems with other technologies that share those same platforms, including applications that benefit and support U.S. national security and economic competitiveness. This issue can impact all phases of the technology life cycle, from development, to production, to use.

THE CHANGING INTERNATIONAL COMPETITIVE ENVIRONMENT

Today, the United States is facing a competitive international environment that is markedly different from the environment that played a large role in shaping the nation's post–World War II competitive and research paradigms, policies, and procedures.

Other countries have been actively challenging the nation's long-standing leadership in fundamental research and technology innovation, most often by emulating the approach taken successfully by the United States: creating world-class R&D environments, developing and attracting talent, and investing in and supporting technology development. Given the strong R&D ecosystems of other countries, it is likely infeasible to prevent competitors from developing many technologies similar to those developed in the United States by restricting access to or the use of those technologies.

As flows of information and people across borders have increased, industrial research and production have become globalized, either because firms have become multinational enterprises with affiliates and customers in many countries or because firms have been increasing offshore research and production. In addition, the United States is no longer one of a small number of countries that produce the highly educated individuals who can drive innovation in emerging technologies. Today, other countries produce more STEM (science, technology, engineering, and mathematics) graduates than does the United States, and have been attracting individuals who were educated and trained in other countries, including the United States, to return to apply their knowledge and skills.

Finally, the United States now faces an adversarial near-peer competitor—China—that over the past two decades has systematically pursued strategies for dominating technology development in key areas. China has been making massive investments in R&D—greater, in some areas, than those of the United States; has a well-educated labor force that is three times larger than that of the United States; and has sought to attract talent from other countries. China also does not play by the same rules as the United States, and makes decisions based on a worldview quite different from that of the United States and its allies. The Chinese government is deeply involved in commercial technology development; research outputs and data from competitors are subject to diversion or theft; foreign participation in the Chinese economy is limited and monitored; technology standards and regulations are managed to advantage domestic technologies; and markets are distorted to advantage domestic companies. China is willing to obtain technology through the acquisition of companies, through foreign-talent programs, and through the theft of intellectual property, and has learned that the United States will often react to such actions by instituting bureaucracies that impede the United States' own capacity to innovate.

The historical approach to protecting technologies in the United States has generally consisted of unilateral reactions to external threats posed by adversaries. Risks in the new global R&D ecosystem cannot be managed effectively in this manner without posing a new risk—that of inadvertently slowing the development and application of technologies and limiting competitive advantage. The U.S. research community has seen an extraordinary increase in the number and complexity of policies, processes, procedures, and requirements governing the conduct of science and technology R&D. That expansion, combined with the increasing array of government stakeholders exercising authority in this realm, has created a set of complex rules that differ markedly across federal agencies with respect to requirements, adoption, and implementation. These rules limit the exchange of ideas, participation by others, and international collaboration, slowing the pace of research and making research environments less attractive to talented people.

PROTECTING OPENNESS WHILE DIFFERENTIATING AMONG RESEARCH ENVIRONMENTS

Limiting the adverse consequences of restrictions on R&D requires defining and maintaining a variety of research environments in which the restrictions being applied are matched to the risks posed by a technology's dissemination. The United States should strive to maximize the amount of work that can be appropriately performed in an open research environment, an approach that will promote U.S. leadership in science and engineering, attract top talent, and enhance discoveries that lead to new technologies. At the same time, this approach recognizes that not all research-related work is appropriate for an open environment. For certain applications, research, development, production, and related activities need to be confined to restricted environments that limit participation, collaboration, the sharing of information, and the dissemination of results to ensure that the knowledge, know-how, production, and use of a technology are limited to those entrusted to use the knowledge and information properly.

Research, training, and teaching conducted in an open environment benefit the United States because they attract research talent, foster creative and innovative conditions for discovery, and speed the development of new ideas and technologies. At the same time, conducting this work in an open environment poses a risk that knowledge, know-how, or results may flow to adversaries as a result of the movement of either information or people. But for an innovation leader, the benefits of openness outweigh the risks for most R&D efforts because the risk of information loss is mitigated by the ability to innovate even newer technologies.

> **Recommendation 1: The President, through an executive order, should clearly reaffirm that it is the policy of the United States that fundamental research, to the maximum extent possible, should remain unrestricted. In addition, the executive order should direct the Office of Science and Technology Policy, in coordination with federal agencies, to define criteria for open and restricted research environments within 120 days of issuance of the executive order. Furthermore, the executive order should direct federal agencies to designate the appropriate environment for work under a grant or contract prior to making the award, and to maximize the amount of sponsored work that can be performed in open research environments. In making this designation, agencies should state clearly that any restrictions or recommended restrictions apply only to the particular research grant or contract being funded, and not universally across the entire institution receiving the funding.**

Just as the United States needs to carefully define criteria and protections for restricted research environments, so, too, must it protect the essential parameters of its open research environments. Once the criteria that characterize an open research environment have been defined, they can be maintained so that any restrictions adopted in the future do not have the effect of slowing R&D. Work that needs to be restricted, whether for commercial or defense reasons, can be performed in other types of environments, such as near-campus federally funded R&D centers or restricted government laboratories, or in collaborative efforts between universities and companies, with the minimum restrictions necessary for protection. If research environments at universities or national laboratories are designated as open, funding agencies can decide a priori what work is to be carried out in those environments as opposed to work that requires a risk acceptance decision.

DEVELOPING AND ATTRACTING TALENT

The strength of the U.S. R&D enterprise is dependent on access to sufficient amounts of high-level talent, both domestic and foreign. Talented scientists, engineers, and innovators are attracted to open and risk-embracing environments where they can pursue promising ideas and be rewarded for their achievements. They may choose to work in environments where technologies are proprietary or classified, but those environments still need to be attractive to smart and ambitious people.

To maintain its leadership in science and technology, the United States will need to continue developing its domestic talent. Yet the United States lags behind other countries in preparing its citizens for participation in technology-intensive areas. Remedying this deficiency remains an urgent public policy objective.

At the same time, however, if the United States is to continue generating more than 20 percent of global gross domestic product (GDP) with only about 4 percent of the world's population, it will have to continue relying on talent from other countries. The nation can no longer be complacent in assuming that the United States is the "default" choice for top global science and engineering talent. Other countries are aggressively competing for top students and STEM professionals, often by emulating the approaches that led to U.S. success in the past. Even in areas in which the United States still enjoys strong advantages, such as attracting top talent for U.S. graduate and postdoctoral training, it lacks coordinated efforts to ensure that those individuals can remain and work in the United States. Imposing excessive restrictions on research environments dissuades talented people from coming to the United States, leading them to find other places to live and work.

> **Recommendation 2: The National Science Foundation (NSF) should fund and coordinate an effort to define those elements of the U.S. innovation system that are essential to developing, attracting, and retaining the top scientific, research, engineering, and innovation talent that is necessary for U.S. leadership in technology innovation. NSF should engage other federal science agencies, universities, research institutions, educators, and research-intensive companies in this effort. The agency should produce a report detailing its findings within 180 days of the start of the effort. Based on those findings, the Office of Science and Technology Policy should coordinate with federal research agencies, the Department of Homeland Security, and the Department of State to develop a national strategy for promoting leadership in science and technology through policies and programs aimed at developing domestic research talent, expanding opportunities for international research collaboration, and attracting and retaining top talent in the United States for training and employment.**

IDENTIFICATION OF STRATEGIC TECHNOLOGIES AND COORDINATED RISK MANAGEMENT

U.S. policy should shift from an approach based on listing "critical" technologies, with associated restrictions, to one based on coordinated risk management. A number of factors—including the rise of platforms; the more diverse set of users and uses of technology (often via platforms); the more diverse set of developers (including commercial and non-U.S. actors); and the more intertwined nature of technologies, markets, and applications—complicate the designation of particular technologies as requiring protection.

The modern features of technology make clear the need for a comprehensive approach to managing the risks associated with strategically important classes of technology development or use. Meeting this need will require developing a risk management approach that begins with identifying which actors using what means are attempting to use a particular technology against U.S. interests or technology leadership, and then defining strategies for addressing the resultant risks. That effort will in turn require expertise that goes beyond the nature of the technology to encompass the plans, actions, capabilities, and intentions of U.S. adversaries and other bad actors, thus involving experts from the intelligence, law enforcement, and national defense communities in addition to agency experts in the technology. This risk management thereby matches particular threats with appropriate strategies for managing them. For a very limited set of technologies, such as those used primarily for national security purposes, risk management may involve routine forms of protection currently employed by federal agencies or commercial enterprises.

> **Recommendation 3: The National Security Council, the National Science and Technology Council, and the National Economic Council should develop and lead an interagency process for identifying and assessing threats or vulnerabilities of strategic significance to U.S. technology leadership and other national interests. For each threat, the process should include developing an associated risk management strategy and evaluation rubric for use by federal agencies in addressing the risk. The execution of these risk management strategies should be coordinated and overseen by the above interagency process to ensure a "whole-of-government" approach. The strategies resulting from this interagency risk management process should be**
>
> - *proactive*, in that they define technology-related threats with national or economic security implications as early in the research and development process as possible;
> - *strategic*, in that they are based on global realities, including the plans, actions, intentions, and capabilities of adversaries, and on reasoned risk acceptance decisions about which technologies must, should, or cannot be protected;
> - *timely*, in that they are based on current understanding of the associated threats and vulnerabilities and are adjusted as required;
> - *integrated*, so that different mechanisms for technology protection, such as export controls, information classification, or decisions by the Committee on Foreign Investment in the United States, are directed and coordinated in such a manner as to effectively reduce or mitigate the risk;

- *adaptive*, with mechanisms for subjecting identified technology areas to regular reviews by integrated expertise in science, technology, and national security;
- *dynamic/repeating*, with a scheduled review to ensure that there have been no changes to the technology, the environment, or the actor(s) that would warrant a change in the threat status; and
- *assessed for adverse effects*, to ensure that they do not result in unnecessary and unintended barriers to U.S. innovation leadership.

This recommendation is not intended to replace the mission agencies' role of identifying risks—or areas of opportunity—to the nation's economic or security interests in accordance with their areas of mission responsibility. Instead, the goal is to provide an effective whole-of-government framework for managing these risks in a coordinated and effective manner.

TAILORED APPROACHES TO THE UNIQUE VULNERABILITIES RESULTING FROM SHARED PLATFORMS

The changing technological landscape has introduced new challenges to management of the risks posed by shared platforms and their supporting ecosystems. Current approaches to risk management assume that each technology is essentially independent of other technologies (regardless of whether they are in fact discrete or separable) and has a single purpose or small set of defined purposes. But approaches to protecting diffuse multipurpose platforms differ from those applicable to discrete, defined-purpose technologies. All countries sharing common platforms also share associated vulnerabilities and risks that affect their national interests. Protecting these national interests typically requires governmental technology policies, such as government involvement in setting standards, regulations, and trade policies.

The committee does not believe that at present, responsibility for identifying and managing the unique risks posed by these shared and powerful platforms is clearly established within the U.S. federal government, at either the federal agency level or the interagency level of the White House. Certain components of risk management suitable for application to platforms do exist in various agencies, but no agency has overall responsibility for coordination of these efforts: Accordingly, the committee believes that the appropriate first steps in identifying strategically important platforms; defining the roles and responsibilities of federal agencies that pertain to those platforms; and developing coordinated risk management strategies covering their development, control, and use should be taken as part of a cabinet-level interagency process.

Recommendation 4: The National Science and Technology Council, the National Security Council, and the National Economic Council

should jointly develop a new policy framework for the identification of strategically important platforms and for the development of coordinated risk management strategies covering their development, control, and use. Elements of this new framework should include

- defining and designating specific technology platforms that are essential to U.S. interests;
- involving the private sector in specifying the technical features and requirements that should be included in platform development, such as performance standards for security, integrity, interoperability, control features, and user controls;
- developing a coherent, whole-of-government strategy for establishing and managing trust relationships among platform developers or users, including international governance mechanisms, use agreements, regulatory approaches, trade agreements, content requirements, and law enforcement cooperation agreements; and
- establishing a range of responses to security or trust problems related to the use of shared platforms, with participating agencies planning for and preparing appropriate "incident response" capabilities.

PROTECTING THE ABILITY TO COMPETE

The openness of the R&D enterprise in the United States has fostered innovation, risk taking, and the incorporation of new ideas into new technologies. It also has attracted the world's best scientists, engineers, and entrepreneurs, whether born and educated in this or other countries, to U.S. universities, companies, and government research organizations.

In today's interdependent, global innovation system, the greatest threat is that the United States will inadvertently weaken its innovation ecosystem while other countries continue to emulate the actions that have historically yielded U.S. advantages in technology development and commercialization. To counter this threat, the United States needs to protect and extend its ability to develop new technologies and apply those technologies to problems in both the military and commercial spheres. Protecting and strengthening this *ability* is vitally more important than protecting specific technologies.

1

Introduction

U.S. leadership in technology innovation is central to the nation's interests, including its security, economic prosperity, and quality of life. To achieve that leadership, both the public and private sectors must weigh competing considerations. The scientific research and development (R&D) that leads to technological advances thrives in conditions of collaboration and the free exchange of information (NASEM, 2018a). The sharing of ideas, participation by others, movement of researchers among institutions and countries, and publication of the ever-growing body of research results (Figure 1-1) spur creativity and speed progress. At the same time, some of those technological advances must be protected to address concerns related to national security or economic competitiveness.

Research environments in which the dissemination of ideas and information is limited are, in general, less productive than open research environments (Felin and Zenger, 2014). The latter environments attract and accept ideas and contributions from multiple sources, whether inside or outside those environments. Subjecting hypotheses, research plans, and draft articles to review by others is a way to identify and correct problems and improve research processes and conclusions. Closed research environments, in contrast, constrain the amount of accumulated knowledge that can be applied to a problem and inhibit the dissemination of new ideas and technologies to others.

CONTEXT FOR THIS STUDY

The openness of its R&D enterprise helped make the United States the world's leader in science and technology after World War II (Kraemer, 2006). It has created a science and technology ecosystem that fosters innovation, risk taking, and the incorporation of new ideas into new technologies. It also has attracted the world's best scientists, engineers, and entrepreneurs, whether born and educated in the United States or in other countries, to U.S. universities, companies, and government research organizations.

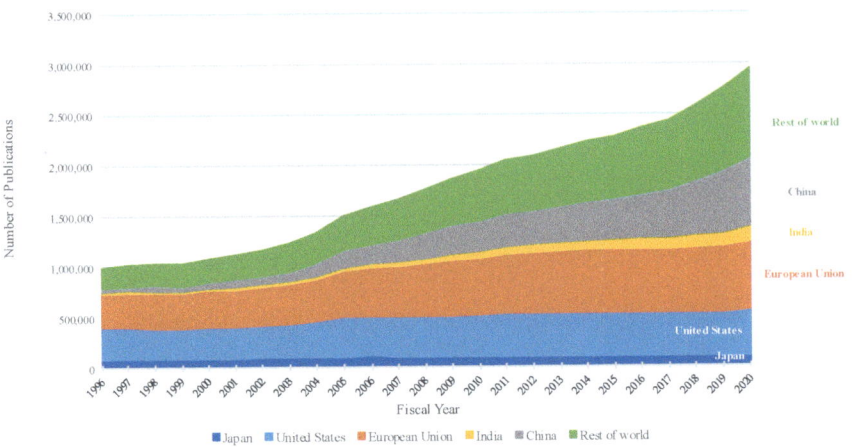

FIGURE 1-1 Science and engineering article publications, by country.
NOTE: Article counts from peer-reviewed journals and conference proceedings in science and engineering fields, assigned to a region/country on a fractional-count basis.
SOURCE: NSB, 2021.

To achieve these national advantages, however, certain technological outputs from innovation processes must be protected from use by adversaries or other forms of dissemination (NRC, 2009). Government, academic, and business leaders uniformly recognize the need to limit the spread of certain kinds of strategically important information (NAS et al., 2009). Information and technologies related to national security are subject to a variety of constraints on accessibility; data on human research subjects may be restricted because of privacy concerns; and companies have incentives to keep some information secret as a potential source of competitive advantage, while government policies often support the protection of proprietary information. Accordingly, the United States has a well-developed system of protections used to limit unauthorized access to technology information, production, or use.

Achieving the right balance between openness and protection can be difficult, especially given recent changes in global capabilities and intentions. Today, the United States is facing a competitive international environment that is markedly different from the environment that played a large role in shaping the nation's post–World War II competitive and research paradigms, policies, and procedures (NSB, 2020a). Once dominant in science, technology, and the industries based on those technologies, the United States and its allies are now competing against other countries with strong science and technology enterprises. One measure of this growing competitiveness, again as demonstrated by Figure 1-1, is the substantial increase in the numbers of scientific publications with authors from China, India, and other countries, such that the United States now accounts for less than one-fifth of total publications.[1] Recognizing the vital contributions that technologies make to their competitiveness, many countries have prioritized investments in their innovation ecosystems. As a result, increasing numbers of technologies, including those vital to military preparedness or economic growth, are being developed and produced in countries outside the United States, often as part of a broader international commercial sector or through global R&D and manufacturing consortia.

Technology itself has changed (NASEM, 2019a). Science, technology, and innovation are much more multidisciplinary, multinational, multipurpose, and interdependent today than in the past, which complicates efforts to protect individual technologies from competitors in either the military or commercial realm. Many new technologies are developed, built, and based on systems of other technologies. As discussed in the next chapter, these "platforms" contribute to the capabilities, scalability, and reduced cost of producing new technologies.[2] They are widely—often globally—shared, are publicly available, and are difficult to protect without adverse impact on all the other technologies that use them.

STUDY PURPOSE, CHARGE, AND APPROACH

To review the protection of technologies that have strategic importance for national security in an era of openness and competition, the Defense Advanced Research Projects Agency (DARPA) and the National Science Foundation (NSF) asked the National Academies of Sciences, Engineering, and Medicine (the National Academies) to convene an ad hoc committee to consider policies and practices related to the production and commercialization of research in domains critical to national security. In its charge (see Box 1-1), the committee was asked to answer three questions:

1. Given today's competitive environment, how should federal R&D funding agencies evaluate or bound the openness of science and encourage transition from idea to commercialization, considering the benefits and drawbacks of specific technology protection and commercialization options?

2. What solutions are required to address market or institutional challenges, if any, related to the production and commercialization of advances discovered in research, particularly those that may have potentially significant impacts on U.S. national security?

---

[1] While publication counts are an inexact measure of R&D outcomes, the bibliometric data used in Figure 1-1 are selected by Elsevier from peer-reviewed scientific and technical journals "based on evaluation by an international group of subject matter experts." The data are further screened by the National Center for Science and Engineering Statistics at the National Science Foundation (NCSES, n.d.).
[2] Notable examples of platforms include app stores, overnight fulfillment and delivery, the internet, genome editing, artificial intelligence, and 3D printing.

> **BOX 1-1**
> **Statement of Task**
>
> An ad hoc committee of the National Academies of Sciences, Engineering, and Medicine will be convened to advise the federal Government on the following questions:
>
> 1. Given today's competitive environment, how should federal R&D funding agencies evaluate or bound the openness of science and encourage transition from idea to commercialization, considering the benefits and drawbacks of specific technology protection and commercialization options?
>
> 2. What solutions are required to address market or institutional challenges, if any, related to the production and commercialization of advances discovered in research, particularly those that may have potentially significant impacts on U.S. national security?
>
> 3. What are the appropriate policy changes related to research, production, commercialization, and technology protection that will help accelerate the marketing/fielding of advances stemming from U.S.-funded research within and to the benefit of the United States, in particular for technologies critical for national security leadership?
>
> These questions will be considered in the context of one or more specific science and technology domains that impact U.S. scientific leadership and national security and will include, at a minimum, synthetic biology, artificial intelligence, and microelectronics (including beyond Moore's Law). Questions of economic competitiveness may be considered when the U.S. economic position is crucial to national security concerns (e.g., maintaining DoD-available supply chains).
>
> The committee will convene a series of meetings, including three workshops, to gather information and consider a broad range of views on the questions above. Drawing from the content of the workshops and other information sources, the committee will review current legal, regulatory, and policy regimes and prepare a consensus report with findings and recommendations.

3. What are the appropriate policy changes related to research, production, commercialization, and technology protection that will help accelerate the marketing/fielding of advances stemming from U.S.-funded research within and to the benefit of the United States, in particular for technologies critical for national security leadership?

The committee addressed these questions in part by examining four specific science and technology domains that have substantial impacts on U.S. scientific leadership, economic growth, and national security: microelectronics (particularly semiconductors), artificial intelligence (AI), synthetic biology, and quantum computing. It also studied aspects of economic competitiveness crucial to national security concerns, such as maintaining supply chains for defense-critical technologies. And the committee reviewed current legal, regulatory, and policy regimes to assess potential changes in policies and practices that would accelerate the commercialization of technologies critical to national security and economic strength. Based on its analysis in these areas, the committee offers in this report recommendations for changes to technology protection policies and practices that reflect the current realities of how technologies are developed and incorporated into new products and processes.

The committee examined technologies related to national security, but the range of such technologies is very broad. Today, military technologies often rely on R&D done in commercial companies, federal laboratories, independent research organizations, and universities. This R&D may occur anywhere in the world, not just in the United States. Furthermore, global power today is closely linked to economic influence, and economic influence is closely linked to technological leadership. Most of the advanced technologies that countries such as China have targeted for leadership are connected in some way to national security, whether because of their direct importance to military technologies or because of their broader effects on economic prosperity. Because of the many interconnections between civilian and defense uses of technologies, this report generally does not use the term "critical" technology, which has come to be associated with technologies necessary for national defense, referring instead to technologies of "strategic importance" to national security and economic competitiveness. Finally, the committee recognized that the world's economies, particularly those of the leading world powers, are much more closely linked than they were in past decades, creating opportunities for leverage but, at the same time, constraints on available policy options.

NEW POLICIES FOR A NEW ERA

One way for the United States to protect strategically important technologies would be to "double down" on existing technology protection approaches, increasing restrictions on the movement of and access to information, technologies, and people. In today's competitive environment, however, the committee believes that such actions would likely damage and slow the rate of innovation in the United States more than it would constrain the advances of other countries. Technologies are developed so rapidly and are so intertwined among sectors and nations that preparing a long list of specific technologies to restrict has little protective value, but instead would throw sand in the gears of the U.S. innovation system.

Similarly, identifying very broad areas of technology or technology platforms as strategic assets that require protection is likely to have a chilling effect on economic growth and, in a dual-use world, on national security. New ideas from basic research and new technologies based on that research are widely available in a world where many other countries have built strong science and technology infrastructures and where research is often conducted by broadly distributed international collaborations. Forcing restrictions on research through the incremental, unilateral imposition of regulations and policies would likely slow research in the United States while having no effect elsewhere, creating more costs than benefits for the U.S. economy and society.[3]

Today, commercial and military advantage goes to those countries that are best able to implement research results most rapidly in the form of useful products, solutions, and processes. These have been, are, and will be the countries with the talent, innovation ecosystems, and resources needed for success.

In today's world, the misappropriation of U.S. technology, while certainly real, is not the greatest threat posed by foreign competitors. Rather, the greatest threat is that the United States will inadvertently weaken its innovation ecosystem while other countries continue to emulate the actions that have historically produced the U.S. advantages in technology development and commercialization. To counter this threat, the United States needs to protect and extend its ability to develop new technologies and apply those technologies to problems in both the military and commercial spheres. Protecting and strengthening this *ability* is vitally more important than protecting specific technologies.

New circumstances call for a pivot from protecting technologies to protecting the advantages of the United States as a leader in technological innovation and development. To maintain its competitiveness, the United States needs to carefully manage its risks in the interdependent, global innovation system that produces today's highly integrated, systems-based technologies, and to do so without trading away its ability to be the "first mover" in developing the disruptive technologies of tomorrow. More broadly, the challenge is to identify both opportunities to maintain or expand global leadership and ways in which the United States can manage national and economic security risks. The United States needs to recognize the elements of its national advantage; strengthen those elements to maintain its leadership in science, technology, and innovation; and identify and close any gaps in its security framework.

STRUCTURE OF THE REPORT

Chapter 2 of this report examines how technologies are developed and commercialized in today's world. New technologies typically consist of tightly connected component technologies that make those new technologies more difficult to control, either by governments or by private industry. As noted earlier, many technologies are developed using powerful platforms that are widely shared among competing nations and function as underlying infrastructure for technology design, production, and use. Successful application of a technology often depends on the speed with which it can be developed and commercialized. The chapter explores these issues by examining four case studies of critical

---

[3] There are a few limited examples of voluntary global agreements to stop dissemination of scientific information, such as the Asilomar agreement.

INTRODUCTION

technology development, focusing on semiconductors, artificial intelligence, synthetic biology, and quantum computing.

Chapter 3 examines the new competitive landscape now confronting the United States and explores the impact of these changes on the nation's approach to managing the risks of strategically important technologies. The systems used by the United States to protect technologies of strategic importance for national security and economic competitiveness are based on assumptions that may no longer apply. Developing a risk management framework to protect U.S. technological advantages in a globally competitive landscape requires a fresh examination and reevaluation of policies and practices long used to protect technology.

Chapter 4 uses the rise of China as a case study of the kinds of competitive challenges the United States is facing today and will face in the future. China has specifically targeted particular technological domains as areas in which it is seeking global leadership. It has subsidized domestic industries; protected industries from international competition; purchased foreign companies in an effort to acquire technology developed abroad; compelled technology transfer in return for market access; and engaged in the theft of intellectual property through such means as espionage, the cyberhacking of businesses and research organizations, and the suborning of research decision-making protocols. At the same time, the United States and China are highly interdependent in science, technology, education, and commerce, which means that risks associated with technology development cannot be eliminated; instead, they must be actively managed.

Chapter 5 synthesizes the evidence and analyses of the previous chapters in the form of 13 findings. As discussed above, U.S. efforts to protect technologies may be counterproductive, slowing and limiting technology development and commercialization when it is the speed with which a technology is developed and applied that often is the critical factor in whether it yields national benefits.

Chapter 6 draws on the content of the previous chapters, on previously published reports, and on the experiences of committee members to present the committee's conclusions and recommendations on how the United States can protect technologies in an era of openness and competition.

2

Changes in Technology Development and Commercialization

The development and commercialization of technologies have changed radically in recent decades (Manyika et al., 2019). Much of the current U.S. approach to managing the risks of technology dissemination was based on the conditions following World War II, when the United States and its allies enjoyed overwhelming leadership in the development of new but discrete and largely well-defined technologies. Today, the research and development (R&D) process that creates new technologies depends increasingly on a level of collaborative, multisectoral, and international collaboration not seen in the past. Many modern technologies are multipurpose, have diffuse origins, and are highly interdependent on other technologies, with owners, users, and stakeholders from multiple countries. Instead of technology emerging from the military to find commercial applications, as in the past, commercial R&D has become the driver of much military technology.

In addition, technology development, production, and commercialization often rely heavily on systems of enabling technology, referred to in this report as "platforms" (see Box 2-1 for more detail). These platforms enable rapid, massive-scale, and lower-cost development by incorporating shareable technology elements into new technology applications. These platforms are often developed and operated by the private sector and have become an essential part of the technology ecosystem.

This chapter explores the changes that have occurred in technology development and commercialization, focusing on both the protection and promotion of technologies of strategic importance to U.S. economic and national security. It also uses case studies of four technologies—microelectronics (particularly semiconductors), artificial intelligence (AI), synthetic biology, and quantum computing—to illustrate these changes and to investigate steps that could be taken to bolster the nation's continued technological leadership.

BOX 2-1
What Is a Platform?

In a technology system, a platform is a system of technologies, processes, information, or services that is used as a base or foundation upon which other end-use technologies, applications, or processes are developed, depend, or are used.

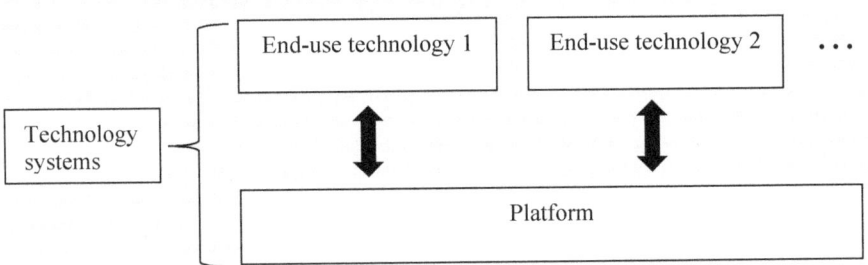

Key Features of a Platform

- **Interoperability:** The rules or features that allow a platform to work together with end-use technologies, applications, or processes that use or are developed on the platform. Interoperability requirements define what universe of end-use technologies, applications, or processes can interface with the platform system. Changes (e.g., upgrades) in either the platform or the end-use technology may affect interoperability between the two. Interoperability is supported by design and standards.
- **Codependency:** The functionality of the end-use technology depends fundamentally on the functionality and features of the platform. These features may be desirable (e.g., capability, power, speed, efficiency, and scale of the process or application), or they may be undesirable or unintended (e.g., vulnerabilities, capacity or reliability limitations, geographic concentration, integrity problems).
- **Sharing:** The platform is widely employed by multiple users, often at global scale. This sharing includes development (shared contribution to the functions of the platform), use, maintenance, and defense of the system from misuse or malfunction. Enabling this shared environment are
  - **standards,** which define interfaces, features, and interoperability of the platform system and between the platform and the end-use technologies; and
  - **governance,** a mechanism for collective decision making and action on shared activities that affect the development, use, function, or features of the platform, including defining relevant standards or rules for the platform and making decisions on operating, maintaining, or upgrading the platform.

Why Emphasize Platforms?

The risks of a vulnerability or threat to a platform cannot be managed using the same approaches used for stand-alone technology systems or products:

- Segmentation of the platform into subplatforms, such as data segmentation, can inhibit the ability of the platform to share functionality or features.
- Modifications to the platform to address security issues affect every user or application on the platform and can affect interoperability.
- Feature changes in the platform can have widespread consequences for a large number of dependent end-use technologies (e.g., bugs creating shared vulnerabilities).
- Sharing can create ambiguity or pose difficult questions on jurisdiction for regulatory, legal, societal, and other key national policies, as well as international governance policies.

Examples of Platforms

- Operating systems (software platforms)
- Internet (hybrid platform for computation and data transport)
- Cloud computing (compute, data, function platforms)
- "Big data" platforms (data platforms)
- Telecommunications platforms (e.g., 5G networks)
- Chip manufacturing fabs (production platform) and the fabless design metaphor
- Shared design "libraries" (design platform)
- Artificial intelligence (e.g., TensorFlow, PyTorch)

HOW TECHNOLOGY DEVELOPMENT AND COMMERCIALIZATION HAVE CHANGED

In the past, technologies were generally optimized for a single, well-defined purpose and had limited dependence on other technologies or systems (Mowery and Rosenberg, 1998). The owners of the technologies were clearly delineated and were usually associated with a single country. Technologies could be characterized as intended for either civilian or military use, and it was possible to limit information flowing between these two spheres. Technologies deemed important for national security reasons typically were protected when they were vital to security interests, and after declassification were released for civilian use. Examples of such technologies include nuclear energy, electronic computing, satellite-based communications (including the Global Positioning System), and the early internet.

Under these circumstances, safeguarding the advantages provided by strategically important technologies generally involved protecting the technologies from disclosure, unwanted production or use, or piracy. Controls were exerted over information, commercialization of technologies, foreign control, trade, technology performance (regarding, e.g., safety), critical materials, fabrication, access, and use. Control often involved labeling technologies as "critical" for either national security or economic reasons, after which an array of protective mechanisms, such as classification or export controls, could be applied to protect them.

While aspects of these circumstances still apply, the world is a very different place today (Dobbs et al., 2015). The globalization of technology development and commercialization has coincided with an extraordinary expansion of international trade (Segal and Gerstel, 2019). As a result, many technologies are developed and produced with contributions from around the world. The semiconductor industry, as described later in this chapter, is an example. Also global are many aspects of biological research, including genomics and disease databases.

One consequence of this global innovation and production environment is the vital role of supply chains that transcend national borders (White House, 2021). Global supply chains have allowed companies to source new technologies from around the world, add needed capability and capacity, cut the costs of production, and pivot rapidly to new products. However, these advantages are accompanied by new risks of lost production and shortages when supply chains are disrupted, as occurred, for example, with personal protective equipment during the COVID-19 pandemic. The loss of production capacity also risks forfeiting process innovations or other forms of innovation that drive technological change (Fuchs, 2014). Offshoring can even be associated with a loss of human capital as innovators leave companies and industries that have moved production elsewhere (Yang et al., 2016). For military applications dependent on these technologies, new risks are associated with ensuring "trust" (e.g., against tampering or alteration) because of the limited availability of domestic manufacturing.

As discussed in detail in Chapter 3, the globalization of knowledge and a great increase in STEM (science, technology, engineering, and mathematics) workforces in other countries have helped create a new environment in which traditional approaches to technology protection are often ineffective and can have substantial drawbacks. The performers of R&D come from the governmental, academic, private for-profit, and even nonprofit sectors, creating a complex combination of actors and nodes both within and across companies. As the rapid development of the COVID vaccines has shown, technologies are developed and move to application much more rapidly and in different ways than in the past. Because research, development, and manufacturing are globally distributed, control of technologies would be ineffective unless extended across multiple geographic and institutional domains. Furthermore, the applications of technologies often are emergent and difficult to foresee in advance. By the time it becomes apparent that a technology should be controlled, it has often spread so widely that controlling it is impossible.

The increasing numbers of well-trained STEM professionals around the world have also changed the development and commercialization of technologies (OECD, 2022). At one time, the United States was among a small number of countries producing highly educated individuals who drove innovation in emerging technologies. (Issues related to human resources are discussed in both of the next two chapters.) Today, however, other countries produce more STEM graduates than does the United States and are building their STEM workforces more rapidly (Buchholz, 2020; McCarthy, 2017).[1] This trend is due in part to individuals who were educated and trained in the United States and returned to their home countries. More recently, other countries have followed the highly successful U.S. model of the post–World War II era and have also greatly strengthened their primary, secondary, and tertiary STEM education systems.

As this globalization of STEM talent has progressed, global technology firms have become dependent on an international supply of scientists, engineers, and other types of technically skilled talent. Supply chains now provide not just raw materials but also intellectual content. Disengaging from these human resources supply chains may be counterproductive or even impossible.

In addition, the interdependence of technologies creates new vulnerabilities. In the case of information technologies, for example, their frequent dependence on access to massive amounts of high-quality data makes them susceptible to efforts to deny that access and distort or destroy valuable information necessary for innovation. The nature of information threats is different in today's digitized and highly interconnected world because the ability to transmit and store information is much greater. Additionally, new technologies can intentionally fabricate or modify data in ways that make verification of authentic content more difficult and create distrust in governments and institutions. As misinformation has become more abundant and more easily accessed, many Americans have come to express distrust in science and technology in general, which can undercut efforts to convey scientific information and build a well-educated STEM workforce.

The Dependence of Military Technologies on the Commercial Sector

Technologies, particularly digital technologies (e.g., communications and networking, autonomy, vision and imaging, AI, and machine learning), are increasingly the bases on which military and national security and conflict depend (Sayler, 2022a). In contrast with the path from military to commercial applications of the past, military technologies have for decades become increasingly dependent on technology development conducted in the commercial sector. In many strategically important technology fields, such as AI, synthetic biology, and microelectronics, the pathway from basic research to application starts with private-sector investments aimed at addressing commercial markets. Decisions made in one domain—regarding technology protections, investments, the focus of work, ethical constraints, and so on—inevitably affect the other.

The case studies examined later in this chapter illustrate this point, but current work on autonomous vehicles provides an immediate and specific example (Lewis, 2021). R&D on autonomous vehicles in academia and the commercial sector has had a major influence on the development of autonomous vehicles for military purposes—initially for logistics vehicles and perhaps later

---

[1] Buchholz (2020) shows that India and China produce more STEM college graduates than the United States.

for combat vehicles. The Defense Advanced Research Projects Agency (DARPA) has had a particular interest in supporting private-sector and academic work on autonomous vehicles through its Autonomous Land Vehicle and Grand Challenge programs. Commercial R&D on autonomous vehicles extends into other domains as well, including AI, robotics, advanced sensing, and digital connectivity. More broadly, this work is helping the United States and other countries build the kinds of innovative, high-technology, and forward-looking economies that will be critical determinants of future national security.

The dependence of the military on commercial technologies illustrates how economic success has become a critical component of national security. Economic strength provides social stability and the long-term ability to fund national defense. It heavily influences relationships with rivals and allies, which is a major component of national security. Consequently, strong relationships with U.S. allies will play an even more important role in technology development. And economic security bolsters the nation's sense of itself as thriving, innovative, and successful.

Key Elements of the U.S. Innovation System

To benefit from new technologies, the United States needs to be a leader in technology innovation and in the incorporation of innovative technologies into new products, processes, and services (NAS et al., 2017). The U.S. innovation system comprises many institutions playing highly specialized roles. These institutions include not just technology companies but also government agencies, government laboratories, universities and other educational institutions, private nonprofit laboratories, research consortia, regulators, standards organizations, trade organizations, and capital markets, along with intellectual protection regimes and other legal frameworks. The network of these institutions is highly decentralized, with significant crossover of ideas and talent but with no policy or central authority responsible for managing the overall enterprise.

These institutions combine in ways that multiply their advantages, but the complexity of the overall system also poses challenges when the need to adapt to changes in technology development arises. The U.S. innovation ecosystem is one of the largest and most mature in the world, but many of its defining features were established when the United States enjoyed overwhelming advantages in technology development and could generally respond to any weakness in this complex and decentralized system by "outinnovating" its competition to regain first-mover advantages. Examples of past U.S. policies aimed at supporting this capacity include major expansions of research funding (e.g., the moon program, nuclear programs, energy research, health sciences research, nanotechnology research, and other programmatic initiatives); incentives to accelerate commercialization of government-funded work (e.g., the Bayh-Dole Act of 1980 [Pub. Law No. 96-517] and the Stevenson-Wydler Technology Innovation Act of 1980 [Pub. Law No. 96-480]); and efforts to "derisk" early-stage R&D (e.g., the National Institute of Standards and Technology's [NIST's] Advanced Technology

Program and its short-lived successor the Technology Innovation Program, grants under the Small Business Innovation Research and Small Business Technology Transfer programs, consortia programs).[2]

The economic and national security challenges facing the United States today from growing international competition in innovation, described in more detail in the next chapter, are the product of a more comprehensive set of circumstances. Addressing these challenges may be more difficult than simply establishing a funding program or policy to accelerate the decentralized U.S. system. Policy responses to the new environment will have to take into account the new challenges stemming from stiff international competition, including areas in which the United States may no longer be the technology leader but may need to "catch up." The current U.S. technology system is deeply interdependent with the systems of other countries, and new risks are emerging in areas, such as foreign direct investment, data access policies, and standards-setting processes, that traditionally have not been subject to focused U.S. policy responses. Furthermore, while competitors have adopted many features of the U.S. approach to technology innovation—including attracting talent, funding research universities, embracing entrepreneurship, and forming risk-taking capital markets—they have not always adopted the U.S. position of government noninterference in the innovation process. The result is a distorted competitive environment that will have to be considered in any policy response from the United States.

The Importance of Platforms

As noted previously, a key characteristic of technology development and commercialization today is the essential role played by platforms. For the purposes of this report, the committee defines a technology platform as a set of integrated technologies, with an associated institutional and human infrastructure, that serves as an essential foundation for the design, development, production, or use of specific technology applications. These platforms are typically multiuse, multipurpose, and multinational systems with many potential applications, often at a global scale, and usually are developed by private-sector actors focused on commercial applications. They can be rapidly scaled and stacked or interconnected, multiplying their effects. They are complex and can change rapidly through innovation and use. Platforms differ from discrete technologies, which typically have a small number of critical components that it may be possible to protect from competitors.

In the field of information technology, a platform is typically a hardware and software environment for building and operating services and processes, such as databases, websites, analytics, or other applications (Sun et al., 2015). This definition can be broadened beyond information technology to include any system used to deliver a discrete service or perform an underlying process, including

---

[2] For more information on these programs, see NASEM (2020a, 2021, 2022) and NRC (2012).

design, development, production, or use. Some examples of platforms are app stores, overnight fulfillment and delivery services, the internet, and the "fabless" semiconductor paradigm, but many other types exist, including design, sensing, production, knowledge, and usage platforms. Many platforms are massive enterprises involving research, production, use, finance, governance, regulation, and trade. At a broader level, they are key components of science and technology ecosystems that accelerate scientific and technological advances and make those advances more widely accessible. They have a significant impact on the features, capabilities, vulnerabilities, fabrication, distribution, and use of technologies.

Powerful platforms enable entrepreneurs to build new products and services quickly, a capability that has created a different competitive dynamic from that of the past. Innovation is not confined within a single technology that one could attempt to protect; rather, it often occurs across an ecosystem of platforms that simultaneously enables new ideas and technologies to be rapidly explored, developed, and commercialized for national security and economic development. Many of these platforms, such as certain communications technologies, are developed in the civilian sector and then adopted for national security applications, illustrating the reversal of the traditional flow of technologies between defense and commercial applications discussed above (Wilson, 2016).

The use of platforms has accelerated the pace of technology development. Platforms such as genome editing, AI, and 3D printing amplify and are amplified by the technologies to which they contribute. In this respect, modern technology has become "autocatalytic," in that the combination of technologies further accelerates the development of those technologies and their cumulative products (Cockburn et al., 2018).

Platforms have attributes that distinguish them from the technologies that countries aimed to protect in the past. These include concerns regarding the trustworthiness of the platform; interconnections, some of which may be hidden or require extreme effort to disentangle; increasing possibilities for unintended consequences as the platforms evolve; and a development speed that outpaces regulatory practices. Also, protecting a diffuse, multipurpose technology platform is much different from protecting a discrete technology with a defined purpose. It is almost impossible to protect a technology that is crucial to national or economic security by controlling only the final application without addressing the contributing features of the platform, because others can usually apply the platform to achieve the same endpoint. Thus, the widespread use of platforms creates new vulnerabilities and risks that affect the national interests of all countries sharing them. But governance of platforms is decentralized and often lags behind the issues raised by the technologies a platform has enabled.

Significant tension can exist among nation-states over control of a platform, as well as over technology upgrades that offer new functionality for shared platforms (Simcoe, 2012). International cooperation may be required in multiple areas, raising questions about the sharing of information and setting of standards. Even within the United States, oversight of the different components of platforms, as well as applications, may fall under multiple agencies (including law enforcement), thereby requiring regulatory harmonization to an extent that has been difficult to achieve in the past.

In addition, the rapid rates of technological change made possible by platforms are poorly matched to legal and regulatory processes. Rapid changes can create sudden disruptions that are poorly understood by policy makers and the public. Small trends or minor developments can suddenly and unexpectedly explode in importance. For example, few people in the 1990s expected the sudden explosive growth of the internet, even though the basis of the internet had been developing since the 1960s. Rapid change also can mean that old and new technologies coexist for extended periods. An example is the simultaneous existence of modern computers and software along with older computers and software with markedly different features, capabilities, and vulnerabilities.

The four technology case studies presented below—microelectronics, AI, synthetic biology, and quantum computing—explore the various influences of platforms on technology development and commercialization.

CASE STUDY: MICROELECTRONICS[3]

Of the four technologies serving as case studies in this chapter, microelectronics has the longest and—thus far, at least—the most consequential history. Semiconductors are absolutely essential to modern industrial and national security activities, and they will continue to be vital to technology development and application. Moreover, their history reveals many lessons that apply more broadly to protecting technologies and platforms critical to national and economic security and accelerating the commercialization of advances stemming from U.S.-funded R&D.

The origins of the transistor and integrated semiconductor chip amply demonstrate the importance of basic research and an ecosystem conducive to technology development. The invention of the transistor in 1947 at Bell Telephone Laboratories in New Jersey, derived from the principles of quantum mechanics and from inspired engineering, was a highlight of 20th-century U.S. innovation (Riordan and Hoddeson, 1997). The development of the integrated semiconductor chip in the 1950s addressed the more immediate need for miniaturization in the U.S. defense industry. The federal government funded up to half of R&D in the nascent semiconductor industry, with that amount tapering off in the 1960s and 1970s as the commercial market expanded. The U.S. government also directly funded the expansion of production capacity in the late 1950s and early 1960s. As prices dropped and capabilities increased, semiconductors became a powerful force multiplier for the U.S. military,

---

[3] This section is based in part on the presentations at a workshop on microelectronics held by the committee on June 10, 2021. An agenda for the workshop and speaker biographies are available at https://www.nationalacademies.org/event/06-10-2021/protecting-critical-technologies-for-national-security-in-an-era-of-openness-and-competition-meeting-4-workshop-on-microelectronics.

contributing to a strategic shift toward electronics-intensive military systems that supported U.S. superiority during the Cold War. By the 1980s, the U.S. semiconductor industry, centered in Silicon Valley south of San Francisco, was vertically integrated and included chip manufacturing.

In the late 1970s and 1980s, Japanese companies, with support from the Japanese government, matched what U.S. companies were spending on R&D in this area and greatly outspent U.S. companies on fabrication capacity. By the mid-1980s, Japan's market share in the global semiconductor industry had surpassed that of U.S. companies. Technology transfer from the United States to Japan initially played an important role in this shift, but Japan also made its own public- and private-sector investments in semiconductor manufacturing technology, which advanced the emergence of Japanese equipment suppliers in the 1970s and 1980s.

Japan's leadership was not to last. The rise of Japanese semiconductor manufacturing led to a policy response from the U.S. government in the 1980s, partly because semiconductors were seen as a national security concern. The Defense Department funded half of an industry-led consortium named SEMATECH (Semiconductor Manufacturing Technology), which partnered with U.S. firms to develop leading-edge manufacturing technology to accelerate the pace of innovation in the U.S. semiconductor industry. In addition, U.S. trade policy placed restrictions on Japan, which opened the door for the emergence of new manufacturers in other countries, especially South Korea and Taiwan. Since then, Japanese semiconductor manufacturing has been in decline, and semiconductor manufacturing is not a large-scale activity in Japan today.

Although the U.S. share of the global semiconductor industry began to grow again in the 1990s, the United States largely did not reenter the commodity semiconductor markets. Instead, it moved toward higher-value products, such as microprocessors, whose value lay more in the design than in the manufacturing technology for the physical device. Today, the U.S. industry includes very few integrated device manufacturers that both design and manufacture their own chips. Most are fabless companies that design custom semiconductors and then outsource the production of the devices to other companies. In shifting toward the R&D-intensive aspects of the business, U.S. semiconductor producers have opted for larger market shares in design tools and smaller shares in the more capital- and labor-intensive segments of the industry. In 2019, North America (primarily the United States) housed 11 percent of global semiconductor fabrication capacity, compared with 40 percent in 1990 (Platzer et al., 2020).

Current State

Today, the semiconductor industry is a global enterprise, with R&D, design, fabrication, assembly, testing, and packaging occurring in many different countries. As just one example, the most important technology for achieving smaller chip dimensions is a device known as a stepper, and a company based in the Netherlands named ASML is the source of the world's leading-edge steppers

(Khan, 2019). Similarly, TSMC in Taiwan is now the world's leading semiconductor manufacturer and provides chip manufacturing and other services for many other companies.

As a result of the globalization of the semiconductor industry, both commercial and military devices have become much more dependent on supply chains that cross national borders. The industry is very complex, a fusion of the public and private sectors with a heterogeneous, mature, and global innovation and manufacturing enterprise. Elements of the innovation ecosystem include R&D, design tools, fabrication facilities of various types, software, security, testing, workforce development, and economics. Semiconductor devices both depend on platforms for their design, manufacture, and use and constitute indispensable platforms used to create other technologies and platforms.

At the same time, the production of semiconductors is concentrated in just a few companies located in Taiwan and Korea. Taiwan, between TSMC, UMC, PSMC, and VIS, accounted for 63 percent of global semiconductor foundry revenue, and Korea, between Samsung and DBHiTek, accounted for 18 percent (Kuo, 2021). In 2020, TSMC alone produced more than half of the world's semiconductors (Kuo, 2021). The market for lithography equipment is similarly concentrated, with the majority of production being performed by ASML, Veeco, and Nikon. In fact, ASML is the sole provider of extreme ultraviolet lithography equipment (Research and Markets, 2021).

The United States has a vital national interest in securing access to leading-edge chip manufacturing processes. Today, however, the United States produces only a small portion of the world's most advanced semiconductors. Most production of these chips occurs in East Asia, and these supplies could be susceptible to disruption because of a trade dispute, natural disaster, global pandemic (as has been the case during COVID-19), or military conflict. For a technology and platform as indispensable as semiconductor chips in both the commercial and defense realms, the concentration of production facilities in any small geographic area or single country poses risks even if that country is an ally. Moreover, such dependence could potentially force the United States to enter into an alliance that might not be beneficial for other reasons or might force the United States to intervene in a conflict.

China's strategy as an industrial competitor poses another risk to continued U.S. access to advanced semiconductors because it is aimed at creating a domestic, vertically integrated semiconductor industry that could challenge the global leadership position of firms in the United States and other countries (Capri, 2020). (Chapter 4 examines the competitive challenge posed by China in greater detail.) The Chinese government is seeking access to foreign capabilities to accelerate the development of the Chinese industry through subsidies, the acquisition of companies and technologies from other countries, and the development and recruitment of human talent. China could severely disrupt the industry through unfair subsidies, dumping, or intellectual property theft. A comparable process occurred with the U.S. photovoltaic industry, which was severely damaged by China's aggressive entry into the market (Fialka, 2016).

Policy Considerations

In an internationally disaggregated production environment (one in which, for example design is separated from fabrication), there can be similar segregation in R&D on next-generation technologies. Thus, offshoring the manufacturing of chips could have a negative impact on the future capacity of the United States to develop next-generation microelectronics fabrication capabilities. Current policies include incentives for the potential "reshoring" of domestic chip manufacturing facilities to the United States through the use of government subsidies.[4] But if these domestic production facilities were limited in use and not part of the global chip manufacturing enterprise, such a policy could increase the cost of maintaining leading-edge technology. The rapid scaling and growth of capabilities in advanced microelectronics are continuing and not slowing. Therefore, any response to these issues must include an element of "moving faster" and restoring U.S. dominance not only in chip design but also in fabrication and integration.

Ensuring security in chip design and manufacturing in a global design and production environment is another challenge. The direct incorporation of advanced-design microelectronics into so many applications makes this a clear case in which many critical technology areas are directly impacted by the integrity, trustworthiness, and function of a device. Compounding the security challenge is that microelectronics are developed and produced through highly specialized, global, and shared design and production systems (that is, through the technology platforms discussed above). The existence of these platforms has allowed various aspects of the process to segment (e.g., into fabless design companies and global foundries), yielding a highly efficient and capable global capacity to design and produce technology products using microelectronics. Not utilizing this global production platform carries significant costs in either capability or financial terms, although an independent domestic supply may help with potential supply chain disruptions.

This global production system poses clear risks, including interference by governments, diversion or sabotage, piracy, industrial espionage, and counterfeiting. Moreover, the concentrations of semiconductor manufacturing and production of lithography tools increases the susceptibility of downstream firms to interruptions in the global supply chain. Technologies or widely adopted and internationally accepted standards, regulations, or norms that could be used to ensure the trustworthiness of fabricated devices do not exist. Alternative approaches include applying trustworthiness frameworks—either through technology or through an international conformance program—to enhance the integrity of electronics.

In the past, the U.S. Department of Defense (DoD) has supported domestic manufacturers to guarantee access to reliable microelectronics important to the national defense. However, this program produces only a small fraction of

---

[4] U.S. Congress, CHIPS Act of 2022, H.R. 4346, Sections 101–107, 117th Congress.

the semiconductors DoD needs, and trusted manufacturers in the United States are hard-pressed to remain at the technological frontier in this arena, which means the chips they produce may be at a technological disadvantage compared with the chips from other sources. Furthermore, malicious software or hardware could be introduced into microelectronics at other times, not just during design or manufacturing. In the past, DoD has certified some domestic manufacturers with trusted supplier status for national defense applications. This program uses a perimeter defense approach and relies on facility and personnel security clearances. Although the trusted supplier model is useful for some defense applications, it does not provide DoD with the most advanced, diverse set of microelectronics as required for national security applications. As a result, DoD has moved toward a policy of assurance based on zero-trust principles instantiated in security standards established in Section 224 of the 2020 National Defense Authorization Act. Recent developments include the Rapid Assured Microelectronics Program, which is aimed at demonstrating measurable security with multiple leading commercial semiconductor companies across the entire life cycle based on commercial best practices, and which is informing the emerging DoD standards. Additionally, programs such as DARPA's Electronics Resurgence Initiative and provisions within the CHIPS Act of 2022 include investments that are intended to accelerate innovation in next-generation microelectronics, as well as to overcome security threats across the entire hardware life cycle.[5]

Policy makers have been debating whether a larger segment of the microelectronics supply chain should reside in the United States to address concerns surrounding manufacturing concentration and the supply chain problems experienced in recent years. In addition to ensuring domestic sources of supply for microelectronics, the concomitant development of human expertise and generation of implicit knowledge could be vital to U.S. economic and national security interests. In modern technology development, including the other case studies examined in this chapter, feedback from the production and use of a technology to R&D activities can be a major source of innovation and advancement. The benefits derived from production and use could be areas in which reliance on international supply chains diminishes capabilities that are foundational to future successes.

---

[5] These security programs include the Automatic Implementation of Secure Silicon (AISS) program, which "aims to ease the burden of developing secure chips. AISS seeks to create a novel, automated chip design flow that will allow security mechanisms to scale consistently with the goals of a chip design. The target design flow will provide a means of rapidly evaluating architectural alternatives that best address the required design and security metrics, as well as varying cost models to optimize the economics versus security trade-off. The target system on chip (SoC) will be automatically generated, integrated, and optimized, and will consist of two partitions—an application specific processor partition and a security partition implementing the on-chip security features. By bringing greater automation to the chip design process, the burden of security inclusion can be profoundly decreased." See https://eri-summit.darpa.mil/eri-programs (accessed August 30, 2022).

The most important factors in maintaining a U.S. semiconductor industry are financial and governmental as well as technological. Creating a better business environment for the industry entails regulations, taxes, investments, and education. People conducting R&D need a strong institutional infrastructure. A level playing field internationally can help foster domestic production.

Managing security risks requires a comprehensive risk management approach, especially as computers become increasingly interconnected. The federal government also has a convening role. Its dialogue with people who are driving an industry, even if the government is a small part of the market, can help in determining what is needed to ensure leadership.

Success in the microelectronics industry, as in other technology industries, depends on access to technical talent. In the United States, access to that talent requires investments in domestic education and research that encourage American students to study STEM subjects. Jobs in the U.S. semiconductor industry are good ones; the U.S. industry directly employed more than 180,000 workers in 2019 at an average annual wage of $166,400—twice the overall average of U.S. manufacturing jobs (Platzer et al., 2020). Maintaining a leading-edge capability in microelectronics also requires retaining and strengthening immigration pathways. Today, two-thirds of graduate students in electrical engineering and computer science globally are from countries other than the United States (NCSES, 2022b). And in 2011, 87 percent of semiconductor patents awarded to top U.S. universities had at least one foreign-born inventor (New American Economy, 2012).

CASE STUDY: ARTIFICIAL INTELLIGENCE[6]

AI denotes machines that perform tasks normally associated with human intelligence, such as driving, spoken-language comprehension, or medical diagnosis. AI is an emerging and highly disruptive technology area with an essentially unlimited range of potential applications. To cite just a few examples, it has been applied to speech recognition, wearable health sensors, cybersecurity, scientific discovery, logistics, games such as chess and Go, medical diagnosis, autonomous vehicles, multiobjective optimization, computer tutoring, robotics, and the modeling of AI systems themselves. AI involves the interaction of technological systems with people, which requires integrating it with fields as diverse as social sciences and psychology. As AI is integrated into current and future technology, it could come to shape every aspect of economic, social, and informational life.

AI also has unlimited military applications (Horowitz et al., 2018). Among these are object recognition and imagery for base defense; precision

---

[6] This section is based in part on the presentations at a workshop on artificial intelligence held by the committee on July 12, 2021. An agenda for the workshop and speaker biographies are available at https://www.nationalacademies.org/event/07-12-2021/protecting-critical-technologies-for-national-security-in-an-era-of-openness-and-competition-meeting-5-workshop-on-artificial-intelligence.

munitions; autonomous systems (including weapons); surveillance; integrated battle-space situational awareness and execution planning; and exosuits that could give soldiers an extra set of "eyes" to monitor their environment and warn them of threats, as well as enhanced physical capabilities in threat situations. Because of these potential military applications, along with its commercial potential, the development and commercialization of AI have garnered the attention of nations around the world. China, for instance, released in 2017 a plan for capturing the global lead in AI development by 2030 (China State Council, 2017).

AI exemplifies the shift today from the historical paradigm of military technology transitioning into the commercial realm, with new commercially developed AI advances often being introduced into the defense regime.[7] This pathway complicates attempts to restrict the technology, since commercial development typically spans countries even if a particular project is occurring within a single company. Furthermore, commercial products that embody AI typically are publicly available, which allows at least part of the technology to spread widely.

AI may be considered a platform technology, or it may be part of a stand-alone technology system. Traditionally, a toolbox of generic AI tools was used to solve specific problems, so that the development of AI components and application domains were distinct and separate. Today, a far more integrated approach to joint R&D is leading to a new discipline that fuses AI and its applications. For this reason, leadership in AI typically requires leadership in the technologies that rely on AI capabilities. In this way, AI has become an accelerant, a catalyst, and a force multiplier across many tasks.

In 2019 the U.S. AI community developed a roadmap for research in the field for the next 20 years (Gil et al., 2019). As this roadmap points out, AI holds the potential to "1) boost health and quality of life, 2) provide lifelong education and training, 3) reinvent business innovation and competitiveness, 4) accelerate scientific discovery and technical innovation, 5) expand evidence-driven social opportunity and policy, and 6) transform national defense and security" (Gil et al., 2019, p. 2). However, realizing such benefits will require fundamental research advances in such key areas as integrated intelligence of modular AI capabilities and skills, sophisticated interaction between humans and machines, and self-aware learning that yields robust and trustworthy AI capabilities. To achieve these advances, the roadmap recommends the creation of a national AI infrastructure marked by open AI platforms and resources, community-driven AI challenges, national AI research centers, and mission-driven AI laboratories.

Current State

The complexity and open-endedness of systems based on AI introduce many risks and vulnerabilities. The performance and capabilities of individual AI

---

[7] The committee notes that some military AI applications, especially those related to target acquisition and attack decision making, do not have commercial-sector parallels.

components have progressed much more rapidly than the ability to integrate them into trustworthy AI systems, and the performance of complex systems built on AI platforms often cannot be characterized a priori. As a result, it is difficult to anticipate failure modes tied to rare inputs. In addition, AI systems can yield biased results because of the data used to train them and system characteristics that are difficult to detect. This lack of interpretability for AI systems leads to significant issues around trust and privacy, pointing to the need for an ethics and legal framework for use of AI for the greater good. Additionally, authoritarian governments around the world may be able to leverage these tools to curtail freedom and democracy. AI-enabled technologies also may exacerbate the spread of misinformation through the transmission of false images and information that are indistinguishable from their true counterparts.[8]

AI has clear national defense–related applications, including autonomous vehicles, facial recognition technology, and maintenance software (GAO, 2022). Moreover, technological advances have led to a proliferation of low-cost AI applications that could lead to increases in the "deployment of AI-empowered drones, cyberattacks, and online information operations" (Kreps, 2021).

This is a complicated backdrop. It is difficult to determine whether AI will remain a set of (many) discrete technology applications or develop into a platform, as discussed earlier, that enables efficient, productive, and capable systems of other technologies. While AI can be integrated with other technologies and platforms to yield new technological capabilities, the combination of deep learning and massive datasets could transform AI into a general-purpose platform that points to new domains of research or areas of technological potential. AI may thus come to serve more broadly as a general method of invention, with profound effects on invention's nature and pace (Cockburn et al., 2018).

AI is already having a profound impact on many aspects of commerce and national security, and is being pursued by scientists, engineers, and innovators globally. Given the breadth of AI and the investments being made privately and by nations around the world, the United States is not in a position to lock this technology away. Attempting to protect AI broadly is unlikely to be possible, given its already ubiquitous nature; runs the significant risk of cutting the United States off from the global focus on and progress already being made in AI itself; and could be detrimental to U.S. interests should AI progress to a general platform.

Policy Considerations

AI-based technologies have the potential to engender profound disruptions in the behavior and capability of a large number of technology markets, military and national security systems, economic conditions, labor

---

[8] Deep fakes and false images use AI to manipulate pictures or videos inexpensively (Langguth et al., 2021).

markets, and systems that support civil societies and social interactions. AI systems may also enhance adversarial capabilities, including crime, counterfeiting, military technology, surveillance systems, and other applications with potentially deleterious consequences.

The potential disruptions resulting from AI raise questions for regulators, policy makers, researchers, and American society about how to oversee its development and use. Protection from AI-related threats may impact technology development, integration into applications, protections on the use of critical data, connections of AI-based systems to physical or logical control systems (such as financial or weapons systems), and user access controls. Both governments and big tech will need to be involved in the oversight of AI-based systems, and they will need to trust each other to move toward a shared understanding of goals and threats and the importance of technology regulation.

Historically, a key factor in the rapid progress of AI research has been the sharing and wide dissemination of fundamental research. As AI is adapted to an increasing number and range of applications, however, the need for restrictions is becoming a more prominent consideration. In 2017, for example, the Trump administration followed the recommendation of the Committee on Foreign Investment in the United States to block a Chinese firm from acquiring a U.S. company that manufactures chips used in AI applications (Hoadley and Sayler, 2020). In the DoD, AI is increasingly being applied to classified information, requiring restrictions on both analyses and the information generated (National Security Commission on Artificial Intelligence, 2021). Even when AI serves as a platform upon which a new application or technology is based, that application or technology may need protection despite an inability to protect the underlying platform.

Interesting developments have occurred involving collaboration between the defense and commercial worlds. Part of the technology—the fundamental or generic AI component—can be developed in an open and unrestricted fashion while some of the more competition-sensitive elements are protected. Thus, mechanisms are needed that allow sharing of data while preserving sensitive or proprietary aspects of the data. Both government and the private sector have responsibilities for establishing such mechanisms. Yet predicting which elements should be protected can be difficult. As the range of commercial and strategic AI applications continues to expand, defining what should be restricted and what should remain open will have a profound impact on competitiveness, on security, and on scientific problems.

Barriers to entry in successful AI technology research vary by application, and with entry depending on other capabilities, such as access to massive sets of high-quality data and high-performance computing platforms. Given that AI is an emerging technology area, the leader in the field will have the greatest opportunities to define the response to various risks. Therefore, moving faster than competitors and adversaries will provide advantages in shaping the technology's development, deployment, and use. A leadership position will also provide more opportunity to shape international systems of governance, including restrictions or limitations on the development or use of the technology.

Other countries are building the large computational and data infrastructures needed to develop AI. They also are supporting broader ecosystems that have the effect of attracting and retaining talent, which will be particularly critical in achieving and maintaining competitiveness in this field. Strong institutions, extensive collaboration, and open research environments are among the factors that will determine success. In contrast, restrictive approaches to technology development have great potential to shift the innovation leadership in AI to countries that are willing to sidestep restrictions to stay ahead. Thus it is important to limit the use of restrictive environments to particular sensitive applications and developments, not to AI more generally.

Because many fundamental scientific breakthroughs in AI are needed, leadership in basic research in this area will be critical. The United States, the United Kingdom, and Canada are among the global leaders in AI research today, while China is investing large sums of money to attain that status. China's applied technology using AI is already very high-quality, as exemplified by its use of speech and facial recognition in state security operations. The Chinese, who operate by a different set of rules from those of researchers elsewhere, are also moving more quickly than is the United States toward the integration of algorithms and data in ways that provide competitive advantage.

Acceleration of AI development includes both R&D capabilities (including research funding and efforts to attract top talent) and commercial and government technology development applications. The development of some AI applications can be extremely expensive and beyond the capabilities of any one country or company, necessitating alliances to share infrastructure and costs. A strategy for both competition and cooperation will need to include everything from research, to the sharing of information, to the standards established for applications.

CASE STUDY: SYNTHETIC BIOLOGY[9]

Although the term "synthetic biology" was first coined in 1912 by the French chemist Stéphane Leduc, its modern conception did not emerge until 1974, when the Polish geneticist Wacław Szybalski described synthetic biology as an emerging phase of molecular biology in which scientists could create new components and add them to existing genomes, or possibly create totally new genomes (NAE and NRC, 2013). Today, although there is no internationally agreed-upon definition for synthetic biology, most people use the term much as

---

[9] This section is based in part on the presentations at a workshop on synthetic biology held by the committee on March 13, 2021. An agenda for the workshop and speaker biographies are available at https://www.nationalacademies.org/event/05-13-2021/protecting-critical-technologies-for-national-security-in-an-era-of-openness-and-competition-meeting-3-workshop-on-synthetic-biology.

Szybalski did—to describe the design and creation of engineered biological systems.

Synthetic biology represents an emerging and disruptive technology with extremely broad potential application areas, including materials science, medicine, manufacturing, agriculture, sensor technology, and human augmentation, among many others (El Karoui et al., 2019). Synthetic biology systems that perform specified functions using DNA or protein sequences—analogous to the use of computer codes to direct functions and outputs—have enabled scientists to turn cells into powerful pieces of machinery. Synthetic biology can be considered a technology "knowledge" or "development" platform, since it will likely have widely shared and distributed features that enable and support other technologies that use synthetic biology. Many experts predict that synthetic biology and biologically based manufacturing will transform the economy and society by replacing products made with traditional materials with products made of sustainable materials.

Applications of synthetic biology raise many questions regarding biosafety, biosecurity, and ethics (Li et al., 2021). Synthetic biology is a dual-use technology with many offensive and defensive military applications. For example, it allows people to develop—either intentionally or unintentionally—pathogens with enhanced transmissibility or lethality, including entirely new kinds of biological agents and toxins (NASEM, 2018b). To illustrate, in 2018 scientists used gene synthesis to reconstruct the genome of the virus that causes horsepox, a relative of smallpox, without having any physical access to the virus and without violating any national or international regulations (Cunningham and Geis, 2020). These kinds of demonstrations suggest that if the United States is to protect itself against the harmful applications of synthetic biology, it will have to establish itself as a leader in the field and work with other nations to develop international norms and regulations regarding its use.

Five major technologies have enabled the rapid development of the synthetic biology field. These technologies—gene sequencing, gene editing, gene arrays, gene synthesis, and single-cell technology—continue to be the foundational tools used to develop new synthetic biology systems. Each of these technologies has undergone advances in efficiency that in some cases rival those characteristic of microelectronics. The Human Genome Project, for example, took 13 years, involved 40 institutions and thousands of people, and cost roughly $3 billion. Today, a single high-throughput gene sequencer could sequence the same genome for less than $1,000—3 million times below the original cost—100,000 times faster than the 40 labs originally involved (Figure 2-1). Similarly, in 2000 a DNA microarray chip could analyze several thousands of pieces of information at roughly $2 per unit of information; today, a modern chip can detect 28 million genetic markers at a cost that is more than a million times lower. These dramatic cost and time reductions are characteristic of platforms, which often offer massive scaling opportunities.

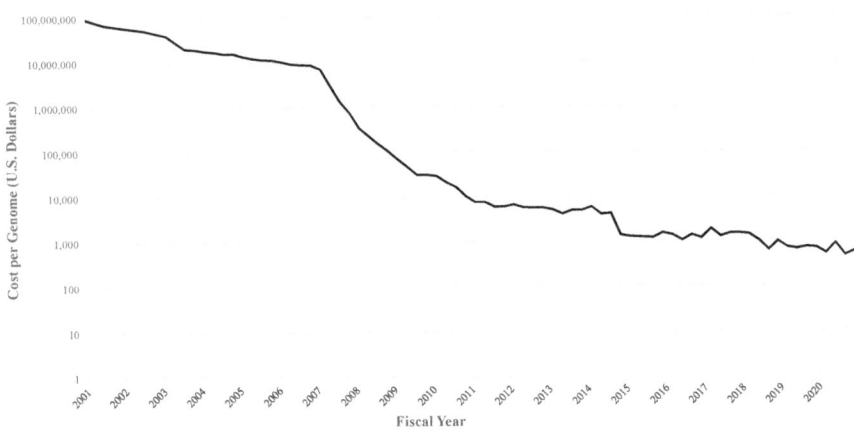

FIGURE 2-1 Cost of human genome sequencing.
SOURCE: Based on data from Wetterstrand (2021).

Although the above five technologies were developed primarily in the academic sector, they were scaled up and applied in the private sector, with engineering playing a pivotal role in their success. Furthermore, as these technologies spread and became more affordable, cost reductions opened up new markets, supporting the bottom-up development of the synthetic biology field. Improvements in these technologies have not only broadened the scope of what synthetic biology can accomplish but also streamlined the experimentation process, requiring less time and effort, and fewer personnel.

Current State

Across its broad applications, synthetic biology aims to design simplified biological components that can be combined to perform specified functions reliably and reproducibly. In this way, it is both generative and open-ended. While synthetic biology is based on biological systems, it represents a novel approach to studying and reimagining those systems using engineering design principles. The interface between engineering and biology makes synthetic biology interconnected, easily automated, flexible, and cost-effective. Moreover, the intersection of AI and synthetic biology will enhance the speed and breadth of synthetic biology's applications. The inherent diversity of life forms and engineering applications means that synthetic biology is centered not around a single goal, but around a conceptual focus on abstraction and simplification. Through synthetic biology, scientists and engineers are building an ensemble of fundamental capacities that will enable humans to partner with living systems in many different ways.

As with AI, a high degree of openness was ingrained in synthetic biology early on. As a result, platforms have diffused across the entire innovation ecosystem, significantly lowering costs. However, openness also creates vulnerabilities. For example, synthetic biology is at risk of being subject to the same mistakes that were made in the development of the internet, whereby security was not built in from the beginning, so that cybersecurity threats are common at both the individual and national levels and are extremely difficult to mitigate, much less eliminate. To set synthetic biology on a different path, protections will need to be incorporated into the design process. Safety measures will need to be automated instead of depending solely on individuals to do the right thing. At the same time, if security constraints are introduced too early in development, it is unclear whether the technology's potential will be fully realized. Fostering good hygiene in the synthetic biology community also means that protections will have to be maintained and updated continually to reflect developments in the field.

Policy Considerations

Many of the technologies that enabled the growth of the synthetic biology field were developed in the United States. As a result, the United States was an early leader in the field, playing a vital role in preliminary research and applications. The United States continues to enjoy a leadership role due in large part to its significant and sustained support for life sciences R&D and a robust investment and commercialization ecosystem. In recent years, however, many countries have identified synthetic biology as a key future player in the biotechnology sector and have begun to invest large amounts of money in its development, threatening U.S. leadership. China in particular has taken a comprehensive and integrated set of actions to gain leadership in synthetic biology. (Chapter 4 examines China's efforts in synthetic biology in more detail.) Barriers to entry in synthetic biology applications are moderately low, and the field has developed in an open research setting in which open dissemination of results and international participation are common. In some ways, synthetic biology is in a position analogous to that of the early internet, with an open research environment, collaboration among trusted parties, and no central governance system.

While organizations such as the International Genetically Engineered Machine (iGEM) Foundation are concerned with education, safety, and security around synthetic biology, the adoption of suggested principles is at the discretion of individual nation-states. Current biotechnology threat frameworks may not apply to synthetic biology applications, and authority is diffuse across multiple federal agencies (NASEM, 2017). The U.S. federal government has yet to develop an overarching funding or governance plan for synthetic biology. Life sciences research is distributed across many agencies and departments of the U.S. government, with no single agency having a primary responsibility for the strength and security of the biotechnology industry. These uncoordinated investments have produced varying degrees of success, raising the question of whether lower-cost, bottom-up strategies in highly emergent systems can drive innovation without the need for top-down planning.

The absence of an international control regime governing synthetic biology creates unique challenges for U.S. national security because of "ethical asymmetry." There exist few effective international frameworks—including standards, regulations, data measurement/transfer, and trade provisions—to support commercial activity in this field. In places such as China, a loose regulatory regime and strong government-led incentives are designed to spur innovation. Thus the United States will need to address the need for security measures while avoiding placing constraints on R&D that competitors are not placing on themselves.

While freely sharing data can create security problems, it can also be hugely beneficial. The open publication of the SARS-CoV-2 sequence in China, for example, allowed research on COVID-19 to proceed rapidly worldwide. To determine how to protect synthetic biology, the U.S. government, academia, and industry will need to share information, ideas, and perspectives to build awareness of security issues, determine best practices, and learn about future and potentially disruptive developments.

Ultimately, the United States remains well positioned to lead the synthetic biology field because of its strength in the life sciences and engineering and its robust startup investment community. Although people disagree on how open the synthetic biology field should be, the use of synthetic biology to improve human lives is likely to create widespread public support for the field. Furthermore, establishing powerful and effective countermeasures for security threats would deter people from using these technologies for harm.

International agreements on standards for design, assembly, data transfer, and data measurement; on regulatory rules; and on the language used could all help advance interdisciplinary and international collaborations that could deliver on the promises of the synthetic biology field. By contrast, absent some form of agreed-upon international standards, many of the products and processes generated by synthetic biology will not translate well to industrial settings, which depend on reproducible processes that are governed by exacting regulatory requirements.

Readying synthetic biology products for the market will require simultaneous progress along many fronts (NAE and NRC, 2013). Existing intellectual property systems will have to be reexamined to determine whether a new national or international intellectual property framework is needed to govern the field's commercialization. Benchmarks will be needed to develop a plan and outline goals for the field. Partnerships with industries will have to be strengthened so that research and manufacturing work hand in hand rather than against one another. Capital investments will need to grow to enable the costly transition from research to production. Proof of effectiveness will be necessary for new products to create trust in these technologies and public support for the field. Concerns about genetically modified organisms will have to be addressed.

These are among the concerns that will have to be addressed in developing a coherent and comprehensive roadmap for the potential future of synthetic biology.

The underlying forms that will define most future activities in synthetic biology are being shaped and crystallized now, both in the United States and worldwide. The United States can only expect to help lead and shape these efforts if it maintains its leadership in the field. To prepare for the future and remain a global leader in the field, the United States will have to make the development of a coherent roadmap and strategy for synthetic biology a high priority (NASEM, 2020b). A roadmap provides targets that incentivize innovation and collaboration, while also providing the flexibility to adapt these targets as the field develops and future breakthroughs reshape the field. Conversely, given the rapid pace of change in synthetic biology, U.S. research leadership in the field could be adversely affected if restrictions were to slow progress.

CASE STUDY: QUANTUM COMPUTING AND QUANTUM INFORMATION SCIENCE

Quantum computing and quantum information and communication technologies represent a future technology area with a very high potential for disruption, including the disruption of existing technologies related to national security. Quantum computers are not the only potential use of quantum control and measurement technologies, even though the field is often discussed using that term.

Quantum computing is the application of quantum laws in the service of computation (NASEM, 2019b). Computation might mean solving a problem that one would ordinarily envision being addressed by a classical computer. Here the term "classical computer" is used to refer to a machine, typically made of solid state bits, that can be idealized as a Turing machine. It is important to keep in mind, however, that classical computers now come as banks of fast processors that work in an almost error-free manner.

Much more progress will be necessary before quantum computers will be able to outpace conventional computers for problems of interest. Still, the key to the possible success of quantum computing is that for certain problems, a quantum computer would require fewer steps to derive a solution relative to any classical computer. This exponential speedup means that even a quantum computer with a slow clock speed can outperform a fast classical computer in some critical applications. The most famous example is the reduced difficulty of factoring very large numbers, which potentially changes the effectiveness of many methods of encryption (NASEM, 2019b). This possibility has attracted the interest of government agencies around the world, especially those concerned with national security.

The forms a large error-corrected quantum computer would take are not clear, and this remains an open area of exciting research. As development progresses, people are hopeful that quantum computers will outperform classical computers in a variety of problem areas, such as quantum chemistry and condensed-matter physics, in which the latter computers have reached their limits.

Current State

At universities, quantum computing and quantum information are seen as important frontier topics of research. The field offers many opportunities for young faculty and postdoctoral researchers, and government funding to pursue these opportunities appears to be readily available. At the same time, private companies have been making extraordinary investments in these new technologies and have been able to lure talent in the field away from academia. Amazon, for example, now invests in both theoretical and experimental quantum science and has constructed an entire building on the Caltech campus devoted to quantum work. Microsoft, Google, and IBM, as well as many other companies, also invest significantly in quantum sciences. In addition, consulting companies are helping traditional companies interface with the new hardware and understand technical issues such as quantum algorithms. For many of these companies, the primary goal is to stay abreast of the field and be ready to take advantage if and when a major breakthrough occurs.

People with the skills required to work in quantum computing and information science are scarce, and the employment market in the field in both industry and academia in the United States, Canada, Israel, Europe, and Asia is extremely favorable. However, the field has not yet advanced to the point at which an established platform exists, and it is possible that this will remain the case, at least in the foreseeable future. Nevertheless, the potential of the science is so extraordinary that the United States needs to be able to compete at the forefront of the field.

Policy Considerations

Uses of quantum computers related to military and national security may emerge through either government or private-sector, commercial efforts. Unlike the other technologies highlighted in these case studies, quantum computing is not a platform technology; instead, it represents a more traditional emergent technology with a high potential for disruption of existing national security–related technologies (along with significant commercial applications). Barriers to entry and the R&D investments needed to create a working quantum computer are very high. For these reasons, in contrast with the other case studies explored in this chapter, much of the existing framework for managing risk applies to this technology.

In the United States, government funding for work by universities and government laboratories in quantum science is augmented by very large-scale commercial efforts. Google, IBM, and the private company Rigetti have prototype quantum computers fabricated with around 100 superconducting qubits, and groups in China are building comparable devices (Parker et al., 2022). The error rates with these devices are sufficiently high that these architectures will not scale up without error correction, so they are not yet of practical use. Nevertheless, these devices have been used to demonstrate "quantum advantage," which means a quantum computer is performing tasks that would essentially take forever on a classical computer.

As in other areas of advanced technology, the continued development of quantum computing and quantum information science will require a collaborative effort among government, academia, and the private sector, which inevitably will include researchers working in multiple countries. A 2016 report from the National Science and Technology Council (NSTC, 2016) identifies five impediments to progress:

- *Institutional boundaries*: Because teams with a diverse range of skills will be needed to make necessary advances, institutional barriers within and among organizations will have to be overcome to encourage research collaborations.
- *Education and workforce training*: Progress will require researchers and research teams with deep expertise not only in quantum mechanics but also in computer science, applied mathematics, electrical engineering, systems engineering, and other fields, which in turn will require new forms of multidisciplinary education and training.
- *Technology and knowledge transfer*: Commercialization of quantum computing applications will require the transfer of knowledge from universities and federal laboratories to the private sector; government programs and policies, such as protection of intellectual property, can enhance this process.
- *Materials and fabrication*: The availability of fabrication capabilities for quantum materials has seen limited progress in some areas, a gap that calls for systems-level engineering that can be carried out, in part, in federal facilities.
- *Level and stability of funding*: Instability of funding caused in part by a lack of coordination among federal agencies has led researchers in the field to pursue alternative careers or opportunities outside the United States where funding is more reliable.

In addition, quantum computing is one of the areas in which commercial advances could be leveraged for military purposes, which complicates the protection of these technologies (Sayler, 2022b).

International competition in areas related to quantum computing is intense, and access to the top talent in the field is a major driver shaping the competitive landscape. First-mover status has given the United States an advantage, and R&D leadership remains a key determinant of success. Maintaining a stable, robust, and well-funded R&D effort is essential to ensure that the United States carries out the first development of this disruptive technology. Because of the very tight global market for R&D experts in quantum technology, the United States needs to anticipate a vigorous effort by competitors and adversaries to attract U.S.-based talent to support their efforts, and to carefully monitor advances in the field. Attraction and retention of key scientists and engineers will need to be a U.S. priority.

IMPLICATIONS FOR POLICY AND PRACTICE

The case studies examined in this chapter lead to several broad conclusions.

First, enabling the ability to develop and apply technologies is generally more important than protecting any specific technology. In particular, providing support for the core institutions that allow technologies to prosper—including educational systems, infrastructure, human resources, and systems of open communications and collaboration—is critical to ensuring national competitiveness.

Second, a major determinant of the nation's future competitive position will be how rapidly new capabilities can be brought to bear on problems. Economic and national security both depend on the ability to quickly incorporate new ideas and technologies into systems that meet important goals. Constraints placed on R&D can hamper the efforts that are essential to remain competitive.

Finally, enabling the ability to develop and use platforms is central to maintaining the competitiveness of the United States. Platforms are the foundational systems that underlie the development and commercialization of new technologies.

The next chapter explores changes in the global competitive landscape. Once dominant in science and technology and the industries based on them, the United States and its allies are now competing against other countries with strong science and technology enterprises. As a result, increasing numbers of technologies, including those vital to military preparedness and economic growth, are being developed and produced in countries outside the United States, often as part of a broader international commercial sector.

3

The New Competitive Landscape

The previous chapter reviewed the dramatic changes that have occurred in how technologies are developed and commercialized in today's global environment. This chapter turns to the competitive landscape—in military, economic, and global ideological terms—that the United States is now facing, one that is vastly different from that of the past. Policies, processes, tools, and approaches developed in previous decades to enhance and protect the nation's economic competitiveness and military security no longer reflect current circumstances and are increasingly counterproductive.

THE COMPETITIVE ENVIRONMENT IN THE 1950–1985 TIMEFRAME

The conditions existing after World War II gave rise to many of the structures and approaches still used today to protect national security. After emerging from the war with an intact industrial base and thriving economy, the United States was the world's dominant economic power. In 1950, with just 5.9 percent of the world's population, the United States accounted for 27 percent of the world's gross domestic product (GDP), a figure that had grown to 40 percent by 1960. By 2020, the United States had 4.2 percent of the world's population and produced about 25 percent of global GDP (Figure 3-1).

The United States also led the world in science and technology after World War II. Following the advice laid out by Vannevar Bush in his 1945 report *Science, the Endless Frontier*, the federal government rapidly increased its funding of research and development (R&D) in the 1950s (Bush, 1945). Federal R&D funding peaked at about 1.8 percent of GDP during the space race of the 1960s, decreased to just above 1 percent through the 1970s and first half of the 1980s, and then began a steady decline until the first decade of the 21st century (Figure 3-2). Meanwhile, starting in the late 1970s, private-sector R&D funding more than doubled as a percentage of GDP, and as a result, total U.S. R&D spending remained relatively stable at about 2.5 percent of GDP.

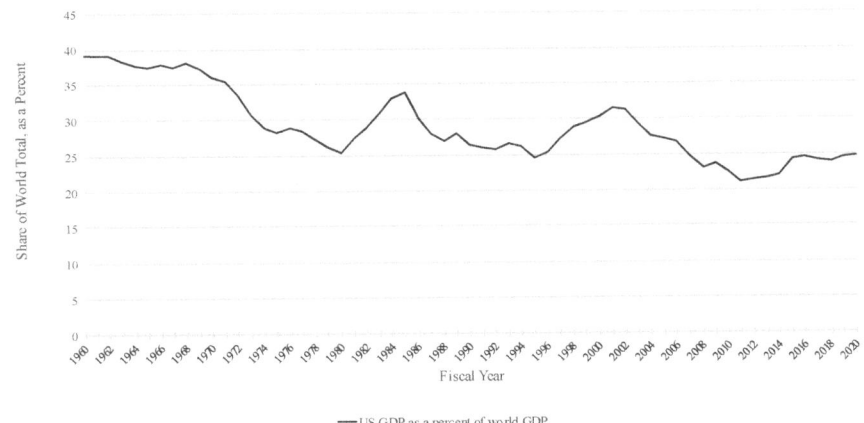

FIGURE 3-1 U.S. gross domestic product (GDP) as a percentage of world GDP, 1960–2020.
SOURCE: Based on data from the World Bank, World Development Indicators (https://databank.worldbank.org/reports.aspx?source=world-development-indicators# [accessed April 20, 2022]).

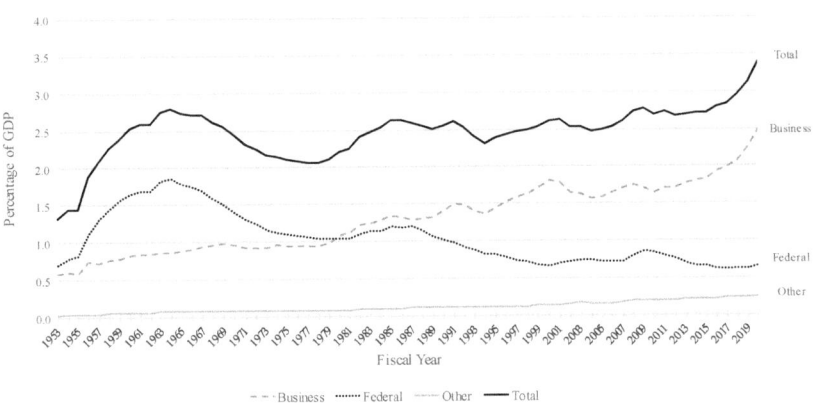

FIGURE 3-2 Research and development expenditure as a share of U.S. gross domestic product (GDP), 1953–2019, by funding source.
NOTE: "Other" includes funding for U.S. R&D by nonfederal government, higher education, and nonprofit organizations.
SOURCE: NCSES, 2022a.

In the decades immediately after World War II, U.S. funding of R&D far outpaced that in other countries. In 1960, U.S. R&D funding accounted for nearly 70 percent of the global total (Scharre and Ainikki, 2020).

Support for science and technology from both the public and private sectors bolstered the nation's military dominance while providing a steady stream of new ideas and technologies that companies could commercialize, often in large industrial R&D centers that produced some of the most groundbreaking advances of the 20th century. Examples of technologies derived in part from defense and nondefense federally funded R&D include digital computing, nuclear power, jet aircraft, microelectronics, spaceflight, and the internet. The U.S. science and technology enterprise was committed to excellence, risk taking, talent acquisition from both domestic and international sources, and broad global engagement in research collaboration. Although patent eligibility and quality vary across countries, U.S. technological leadership for much of the past century was reflected in the number of patents filed with the U.S. Patent and Trademark Office (USPTO) (see Figure 3-3) and the number of scientific publications with at least one U.S. author.

Accompanying the growth in federal R&D funding in the 1950s and early 1960s, support for higher education through such mechanisms as the Servicemen's Readjustment Act of 1944 (the "G.I. Bill") and an increase in the number of jobs requiring postsecondary education led to a rapid expansion of higher education in the United States. Enrollment in colleges and universities rose

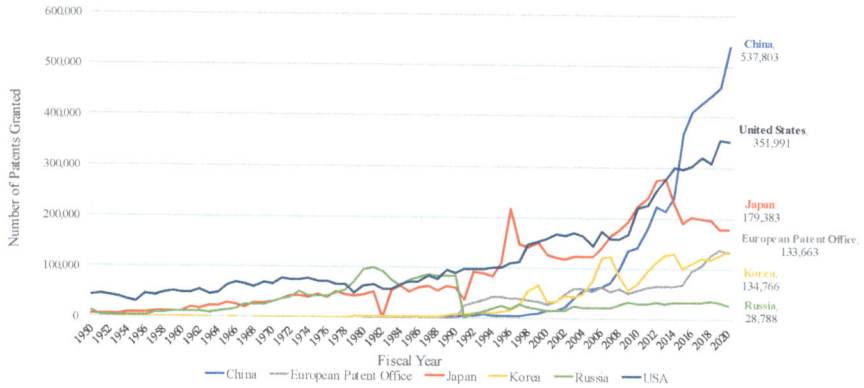

FIGURE 3-3 Number of patents granted, 1950–2020, by filing office.
NOTE: China includes Hong Kong and Macao SAR.
SOURCE: Based on data from the World Intellectual Property Organization (WIPO) Statistics Database (https://www3.wipo.int/ipstats/index.htm?tab=patent [accessed June 6, 2022]).

from less than 2.5 million in 1950 (representing 10 percent of the U.S. population of 15- to 24-year-olds) to more than 8.5 million by 1970 (23 percent of 15- to 24-year-olds), and to almost 14 million by 1990 (37 percent of 15- to 24-year-olds) (Snyder, 1993). Although the number of 2- and 4-year colleges and universities in the United States roughly doubled over these four decades, these increased enrollments were accommodated to a greater extent by an expansion of existing institutions. Many institutions also strengthened their programs in STEM (science, technology, engineering, and mathematics) to meet the needs of their burgeoning student populations and of the workplace.

The dominance of the United States and its allies in the development and commercialization of new technologies gave the nation numerous "first-mover" advantages over other competitors. U.S.-based companies were often the first to develop new technologies and deploy them in the market, and in so doing were able to shape market conditions. As a result, the United States enjoyed advantages in developing the standards that defined new technologies (especially platforms), in shaping the user base for those technologies, and in defining the regulatory frameworks to support them. Being the first to specify new applications of these technologies enabled the United States to identify new markets that attracted investment and talent, defined supplier relationships, and shaped trade relationships. Even when the production of more mature technologies moved to overseas locations, the United States was generally able to respond by developing the next new technology area, where it would again enjoy these first-mover advantages.

Economic growth driven by public- and private-sector investments in R&D, a world-class university-based educational and research system that attracted talent from around the world, and an innovation environment that both celebrated and rewarded risk taking established the United States as the world's leading economic superpower in the 1950–1985 period. Ownership of intellectual property and the broad availability of U.S.-based investment capital contributed to successive "new economies"—consumer electronics, air flight, digital information technologies, health care, and so on—that defined the postwar era. By the 1970s, a vibrant research and innovation ecosystem, world-class universities serving a broad population, a strong national laboratory system, substantial and consistent government research budgets, highly visible leadership on the world's scientific and standards stages, and an aura of energy and invincibility combined to make the United States a mecca for international students. Just as Europe was once a top destination for U.S. graduate students and postdoctoral fellows, the United States became a leading destination for talented students, many of whom entered the U.S. workforce after graduation.

THE COLD WAR NATIONAL SECURITY COMPETITION

The decades from 1950 to 1990 were also characterized by a bipolar military competition that pitted the United States and its allies against the Soviet Union and its allies. Military competitiveness was maintained through national investments in scientific research, technology development, weapons, and personnel. Individual technologies, such as nuclear weapons, stealth aircraft, cruise missiles and precision-guidance munitions, were a major basis of competition, creating the so-called offset paradigm of national defense in which science and technology leadership and industrial might more than compensated for the numerical personnel advantages of adversaries. Military technologies were pursued through large-scale acquisition processes, and the private-sector technology connection to defense was through the defense industrial base. And while many factors ultimately contributed to the collapse of the Soviet Union and the ultimate "winning" of the Cold War by the United States, U.S. primacy in science, technology, and innovation was decisive. A primary argument of this report is that while today's competitive environment is substantially different from that during the Cold War, leadership in science and technology remains the most important capability for the nation to leverage.

In the competitive environment of the Cold War, the United States could outinnovate and outspend its competitors to maintain its scientific, technological, commercial, and military advantages. During this "rising tide lifts all boats" era, the United States could remain well ahead of other countries while being open and allowing others (especially U.S. allies) to obtain the products of its R&D enterprise. This stance also enabled the United States to take maximum advantage of global talent and expertise. The technologies that drove military competitiveness were relatively distinct from those that drove commercial products and markets, and their dissemination could be clearly and vigorously controlled without substantially slowing the development of commercial technologies. The strong separation between military and commercial technologies also meant that foreign researchers could be provided with abundant opportunities to learn, create, and develop technologies in the United States and then stay to contribute their expertise, innovation, and energy to U.S. economic growth without raising national security concerns.

Over time, many scientists and engineers from other countries became U.S. citizens and moved into corporate and academic leadership roles, thus bringing greater diversity to U.S. institutions. U.S. researchers and research institutions also collaborated with researchers in other countries and encouraged foreign researchers to use U.S. facilities.

THE RESULTING POLICY LANDSCAPE[1]

During the Cold War, protecting the many U.S. advantages associated with technology leadership essentially was distilled down to a range of policies aimed at protecting specific "critical technologies" from unauthorized disclosure, production, or use by adversaries. During this period, the risks associated with

---

[1] This section is based in part on a presentation to the committee on August 10, 2021, by Gerald L. Epstein, distinguished research fellow at the Center for the Study of Weapons of Mass Destruction, National Defense University.

these discrete technologies, typically with single or limited uses that were clearly related to national or economic security, could be managed through the application of controls or restrictions on their development, manufacturing, use, acquisition, or trade.

However, it was the assumption that the United States enjoyed overwhelming advantages as a first-mover technology innovator that justified this policy approach based solely on the restriction of specific outputs of that enterprise. Furthermore, the R&D advantages of the United States over its adversaries enabled it to tolerate any potential downstream risks associated with these restrictions because they were unlikely to have a negative effect on the nation's leadership position.[2] It then follows that if U.S. dominance in the development of new technologies can no longer be assumed, this policy approach may not adequately protect U.S. interests.

National Security Decision Directive 189

National Security Decision Directive 189 (NSDD-189), entitled "National Policy on the Transfer of Scientific, Technical and Engineering Information" and issued on September 21, 1985, was a central policy-defining document governing federal R&D restrictions during the Cold War (White House, 1985). Reaffirmed by President George W. Bush, NSDD-189 reflects the philosophy underlying how the Unites States views the value, role, and performance of fundamental research and the balance between the risks of openness and restriction. The directive establishes that "the free exchange of ideas" is "vital" to the strength of American science. It defines "fundamental research" as "basic and applied research in science and engineering, the results of which ordinarily are published and shared broadly within the scientific community, as distinguished from proprietary research and from industrial development, design, production, and product utilization, the results of which ordinarily are restricted for proprietary or national security reasons." It then states:

> It is the policy of this Administration that, to the maximum extent possible, the products of fundamental research remain unrestricted. It is also the policy of this Administration that, where the national security requires control, the mechanism for control of information generated during federally-funded fundamental research in science, technology and engineering at colleges, universities and laboratories is classification. Each federal government agency is responsible for: a) determining whether classification is appropriate prior to the award of a research grant, contract, or cooperative agreement and, if so, controlling the research results through standard classification procedures; b)

---

[2] A secondary assumption was that the United States had sufficient leverage to convince other potential countries with militarily sensitive technology to undertake similar controls on exports to potential adversaries and not to allow third-party transshipment (see, e.g., Gompert and Kugler, 1996).

periodically reviewing all research grants, contracts, or cooperative agreements for potential classification. No restrictions may be placed upon the conduct or reporting of federally-funded fundamental research that has not received national security classification, except as provided in applicable U.S. Statutes.

At its core, NSDD-189 addresses many of the issues the committee has been asked to review. It recognizes the profound value of fundamental research and observes that such research provides the best value to society if its results are communicated broadly, rapidly, and without encumbrance. It further recognizes that some areas of research are vital to the U.S. national security and that such areas need to be protected through simple and limited mechanisms. As the Center for Strategic and International Studies has stated, "This Directive does not assert that the open dissemination of unclassified research is without risk. Rather, it says that openness in research is so important to our own security—and to other key national objectives—that it warrants the risk that our adversaries may benefit from scientific openness as well" (Commission on Scientific Communication and National Security, 2005).

At the same time, it should be recognized that NSDD-189 reflects the era of U.S. dominance in science and technology during which it was issued. It is focused on restrictions on the products or results of research rather than the research process itself and on restrictions on the production, sale, trade, or use of technologies. It is informed by risk acceptance position that the costs of losing some information of commercial or national security importance to other countries are outweighed by the benefits of openness. The policy simply does not envision the potential need to protect the U.S. position as the leader in R&D and the development of new technologies or the underlying conditions that are responsible for the nation's leadership. As a result, the directive does not address the need to protect access to top talent or to preserve open environments that foster disruptive discoveries. While NSDD-189 remains an important statement of principle for U.S. policy, then, its sole focus on protection of the information and technology outputs of the R&D process through such restrictions as classification limits its usefulness in a more competitive global environment. The following subsections briefly review some of the most common restrictions used to manage technology- or research-related risks.

Classification

Consistent with the priority accorded open fundamental research by NSDD-189, the United States maintained a relatively simple formal mechanism for classifying research results. Almost all security classifications are assigned under an executive order rather than under statute. With rare exceptions, information can be classified only if it is owned by, is produced by or for, or is under the control of the U.S. government, a provision that greatly restricts the government's ability to classify information in, for example, an individual researcher's publications.[3]

Under an executive order issued during the Obama administration, information regarding "scientific, technological, or economic matters relating to national security" may be classified, but "basic scientific research information not clearly related to national security shall not be classified."[4] Classification of information on national security grounds prohibits its access by anyone without a government-issued security clearance and a demonstrated "need to know."

Federal Contracts and Grants

The terms and conditions of federal grants constitute another mechanism by which the government can place controls on research. Provisions in grants and contracts may designate results as requiring protection, restrict publication, provide for advance government review, or require approval of individuals performing the research. Although terms and conditions vary by federal agency, for example, grants and contracts typically encourage principal investigators working on unclassified projects who are concerned that their research results may be classifiable to bring those concerns promptly to the attention of the funding agency. However, researchers are not always aware of these requirements or in the best position to identify the potential uses of their research that would warrant classification.

The terms and conditions of federal funding typically require disclosure of foreign financial conflicts of interest and commitments and of any foreign affiliation. As discussed below, the White House has provided guidance designed to clarify the implementation requirements of a National Security Presidential Memorandum issued in early 2021.

Controlled Unclassified Information

Some 70 statutes provide for control of unclassified information for which some level of protection from unauthorized access is deemed necessary. In the past, this information was labeled in a variety of ways, such as "sensitive but unclassified" or "for official use only," and federal agencies developed their own procedures for protecting such information.

In 2008, the Bush administration issued a memorandum entitled "Designation and Sharing of Controlled Unclassified Information (CUI)," aimed at standardizing the terms and practices used by departments and agencies in designating research results as requiring some form of protection. Nonetheless,

---

[3] A major exception is research related to the development of nuclear weapons. According to the Atomic Energy Act of 1946, "The term 'restricted data' means all data concerning (1) design, manufacture, or utilization of atomic weapons" (Pub. Law 83-703, Section 11.y).

[4] The text of the executive order is available at https://www.archives.gov/isoo/policy-documents/cnsi-eo.html.

policies and practices still differ across departments and agencies and even differ from program manager to program manager, as described later in this chapter.

Export Controls

Another mechanism by which the government can control knowledge or information relevant to innovation involves export controls. The 1949 Export Control Act gave the U.S. government the legal authority to restrict exports to Soviet bloc countries, and an international committee, the Coordinating Committee for Multilateral Export Controls, established by the Western bloc in 1950, developed lists of strategic technologies and materials subject to controls.[5] Under the export control system, licenses are required to export certain items to certain destinations, with the requirements depending on both the item and the destination. Exports of nonpublic technical data associated with listed items can also be controlled.

Transfers of controlled information to a foreign national within the United States are also deemed to be exports. Such transfers can require a license even though the information never crosses national boundaries. Employers must have a license to share controlled information with foreign nationals in their employ.

Publication of fundamental research[6] is generally exempt from export controls. Thus, the publication of fundamental research information is generally not considered a licensable act within the export control system. Voluntary government security review, or mandatory review of government-funded research, does not change the export control status of fundamental research. However, material redacted from publication may become subject to export controls.

Controls on Foreign Investment

The President has the authority to block investments by foreign entities in U.S. companies or real estate when those investments might impair national security—for example, by offering foreign access to or control of technologies, data, or other capabilities important to military or other national security systems. Such investments are reviewed and approved by the Committee on Foreign Investment in the United States (CFIUS), which was created in 1975 in response

---

[5] The Coordinating Committee for Multilateral Export Controls was later largely succeeded in the 1990s by the Wassenaar Arrangement; 42 nations now participate in these voluntary export controls of conventional weapons and dual-use goods and technologies. See https://www.wassenaar.org/about-us/#faq (accessed August 30, 2022).

[6] "Fundamental research" is defined as basic and applied research in science and engineering, the results of which ordinarily are published and shared broadly within the scientific community, as distinguished from proprietary research and from industrial development, design, production, and product utilization, the results of which ordinarily are restricted for proprietary or national security reasons (see NSDD-189 [White House, 1985]).

to concerns about investments in the United States by Middle Eastern countries. Because the nation has an interest in encouraging foreign investment, CFIUS may seek some sort of mitigation that would enable a sale to occur, such as spinning off a U.S. division or establishing a control authority mechanism. Recent changes to CFIUS are discussed later in this chapter.

Other Restrictions

Other restrictions also exist. In the life sciences, policies on "dual-use research of concern" establish mechanisms for identifying security risks that the research may pose and considering mitigation approaches.[7] These mechanisms are required as a term and condition of U.S. funding and do not directly affect privately funded activities not conducted at federally funded institutions. However, many private entities have established processes that parallel those of the government.

A wide range of policies in areas unrelated to security—such as safety, ethics, use of animals, use of human subjects, and environmental protection—can constrain research and international collaboration and affect innovation. Such policies serve social objectives that may be judged more important than any competitive disadvantage they may produce.

Standards, Trade, and Other Controls

Technical standards representing agreement on shared properties or features of technologies or the processes that produce them play a major role in shaping which technology features will be acceptable in the market (because they conform to the standards). Therefore, standards play a significant role in defining a technology, particularly any widely shared system technology or platform. Most standards are developed by standards-setting organizations, which may be international, governmental, or private sector. The position of the United States has been that federal use of standards—for example, for regulation or procurements—should be based on a voluntary consensus standards-setting process and that the government should participate in that process (OMB, 1998). This bias toward using industry-developed standards was shaped by the dominant role of U.S. technology companies in shaping the technology landscape. This "hands-off" approach is starkly different from the approach of some of the United States' national competitors, for which standards setting and recognition are government functions.

As a control, standards play an indirect role. By defining "preferred" technologies as those conforming to standards (for example, a standard data format or communication protocol), they can affect market access by other technologies or distort competition unless they are accessible to all. Similarly,

---

[7] For the National Institutes of Health, these mechanisms are described at https://oir.nih.gov/sourcebook/ethical-conduct/special-research-considerations/dual-use-research.

some countries require disclosure (for example, of source code or of other proprietary details) of technologies incorporated by a standard. Standards also play a key role as a basis for regulation or trade, and in this way can create (or eliminate) regulatory barriers, or technical barriers to trade.

TODAY'S COMPETITIVE LANDSCAPE

Since the end of the Cold War, and at an accelerating pace since the beginning of the 21st century, the U.S. science, technology, and economic landscape has changed dramatically. Major competitive shifts have occurred in the global nature of science and technology and in the commitments and capabilities of global allies, partners, and adversaries. These fundamental changes necessitate a reexamination of the foundational bases for and the processes used to evaluate how best to protect and enable national security and economic competitiveness.

This section focuses primarily on the research and technology elements of the changing competitive landscape. Numerous other tools, processes, and mechanisms also are used to build, enhance, and sustain a nation's economic and national security competitive position. Two used effectively by the United States—participation and leadership in setting international standards, and CFIUS—are described in the previous section. Several others are briefly noted here, including international agreements in trade, research, intellectual property protection, and security (e.g., weapons nonproliferation); a welcoming immigration system for top global talent; and tax and investment structures that effectively lubricate the movement of technologies to business value.

Global competition in science and technology grew as other countries witnessed the success of the United States in science, technology, and engineering and the profound importance of its higher education research institutions in enabling that success. In response, many adopted elements of this "American model" and began investing more in R&D (see Figure 3-4). According to one recent estimate, Chinese investments in R&D will surpass those of the United States by 2025 (Chik, 2021). From a high of 69 percent of the R&D performed worldwide in 1960, U.S. R&D spending fell to 30 percent of the global total in 2019 (Sargent, 2021).

Increases in R&D funding have helped spur increases in the number of patents granted and the amount of venture capital (VC) invested. As shown in Figure 3-5, global VC investment skyrocketed in 2021 amid the COVID-19 pandemic. However, the U.S. share of global VC deal value has remained below 50 percent since 2015. The reason is that, although the United States still leads all other countries in VC investments (see Figure 3-6 for 2021 breakdown), VC investment in China and India has been growing rapidly.

Other countries have also greatly increased the size of their tertiary education systems and the number of students graduating with undergraduate and

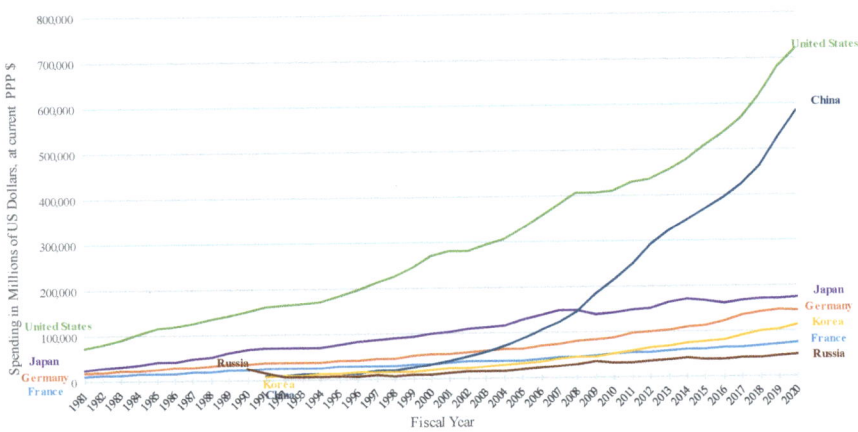

FIGURE 3-4 Public and private spending on research and development, 1981–2020, by country (top seven spenders).
NOTE: PPP = Purchasing power parity.
SOURCE: Based on data from OECD, 2022.

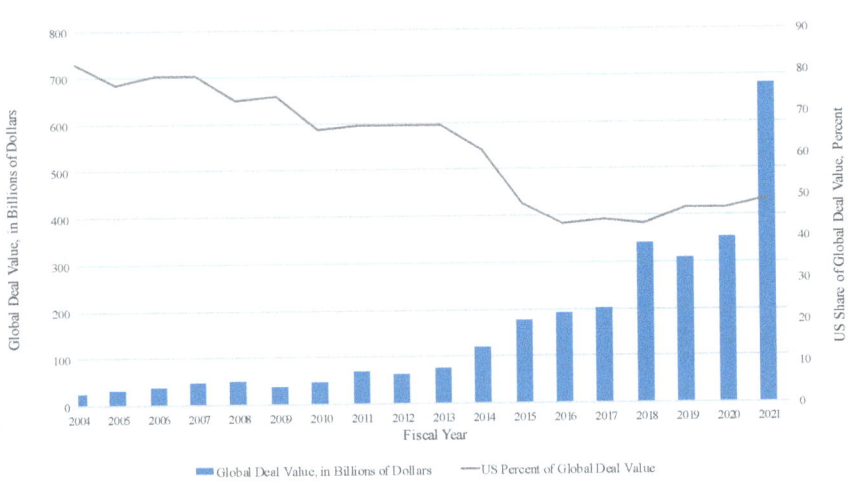

FIGURE 3-5 Global value of venture capital deals in billions of U.S. dollars, and the U.S. share of global deal value, 2001–2021.
SOURCE: Based on data from NVCA, 2022.

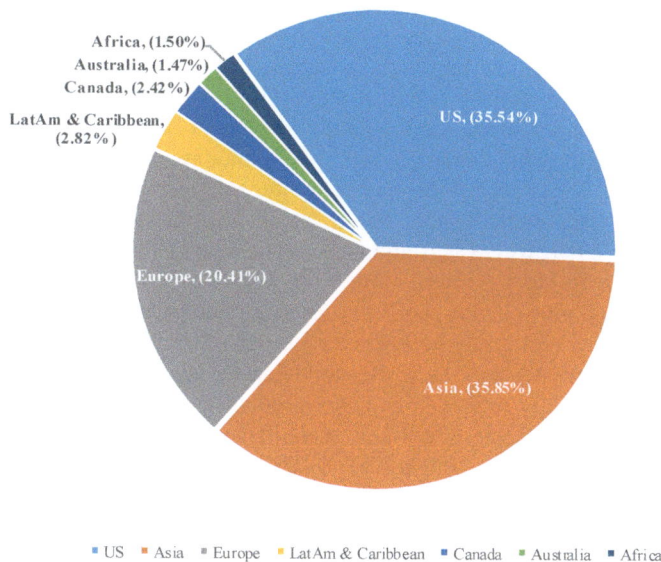

FIGURE 3-6 Share of venture capital deals in 2021.
SOURCE: Based on data from CB Insights, 2022.

graduate degrees in STEM fields. In 2018, approximately 2.3 million students in India and 1.8 million in China graduated with first university degrees in science and engineering, compared with about 810,000 in the United States (NSB, 2022b). China and the United States now award about the same number of doctoral degrees in science and engineering, and many of those degrees in the United States go to international students on temporary visas (as discussed below and in the next chapter). By supporting domestic programs in STEM education, foreign governments are pursuing sustained, systematic strategies for reducing U.S. advantages in innovation and production.

One result of this emphasis on R&D and STEM education by other countries is that scientific research and technology development have become much more internationally integrated. Academic researchers collaborate and access sponsors globally, communication is instantaneous because of new digital communication technologies, and collaborators no longer must meet in person. As one measure of this globalization of research, the percentage of worldwide science and engineering articles produced with authors from research institutions in at least two countries rose from 18 percent to 23 percent between 2010 and 2020. Among articles with a U.S. author, about 40 percent have authors from multiple countries (NSB, 2019).

Another measure of the globalization of science and technology is the number of international students who come to U.S. colleges and universities, often to study in STEM fields. The share of foreign-born graduate students and postdocs enrolled in U.S. universities has risen substantially since 1980, although that share recently dropped, likely as a result of COVID-19 (Figure 3-7) (Channa, 2021). Students on temporary visas earned about 36 percent of the master's degrees in science and engineering awarded in 2019, including 50 percent of those in engineering and 75 percent of those in computer sciences (NSB, 2022a). Students on temporary visas earned about one-third of doctoral degrees in science and engineering in 2019, which was about the same proportion as in 2011. In 2019, temporary visa holders earned more than half of U.S. doctoral degrees in economics, computer sciences, engineering, and mathematics and statistics. After graduation, many recipients of bachelor's, master's, and doctoral degrees do remain in the United States to work, contributing substantially to the nation's science and technology enterprise and entrepreneurial vigor. Additionally, foreign scientists and engineers are more likely than their native counterparts to start new high-tech and high-growth companies (see, e.g., Azoulay et al., 2022; Kahn et al., 2017), although recent evidence suggests that U.S. immigration policies may be inhibiting the ability of foreign students to found new companies or work in startups (Roach and Skrentny, 2019; Roach et al., 2019).

Federal policies imposing restrictions on international students may be one reason for the plateauing of international graduate students studying in U.S. institutions. Another may be that other countries have instituted or expanded programs to retain talented students in science and engineering and attract graduates of U.S. universities back to their home countries. The United States is no longer the only aspirant or destination for global talent. Immigration systems in other English-speaking countries, such as the United Kingdom, Australia, and Canada, are more flexible and more focused on skills-based needs compared with the U.S. immigration system (NASEM, 2015). High-skilled immigration is growing more rapidly in other Organisation for Economic Co-operation and Development (OECD) countries than in the United States, and the U.S. share of international students has been declining (NASEM, 2015).[8] Other countries have launched initiatives to recruit talent, including students who have been trained in the United States. While recent U.S. immigration reforms have allowed foreign students to work longer (through optional practical training) after completing their degrees, the number of temporary work visas is capped, even for those occupations in which workers are in high demand.

Reflecting trends in higher education, technology-based companies have become more international since the 1980s, in part because of the globalization of science and technology, the development of digital communications, and the

---

[8] Rising costs of U.S. higher education, visa delays and denials, and expanded opportunities in other countries are some of the reasons listed for the decline in attractiveness for international students (Israel and Batalova, 2021).

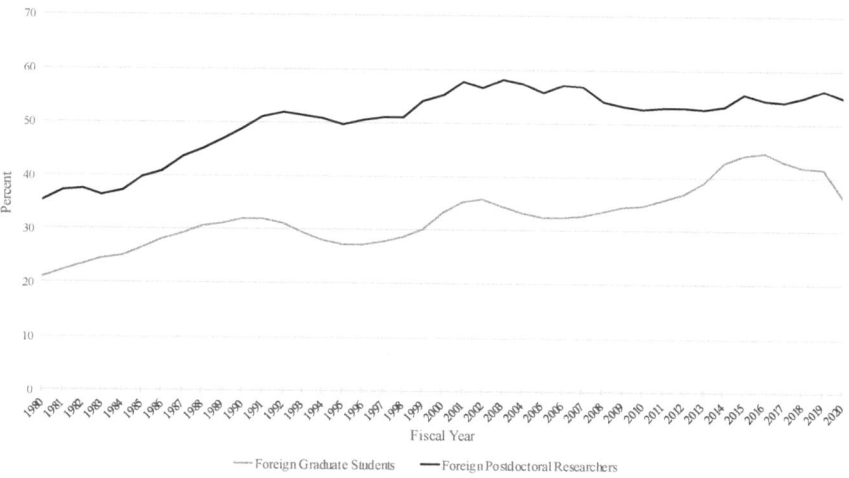

FIGURE 3-7 Percentage of foreign graduate students (master's and doctoral programs) and postdoctoral researchers in science and engineering in the United States, 1980–2020.
SOURCE: Based on data from NCSES (2022b).

establishment of global supply chains. These companies do research globally, seek to tap into local innovation systems and the products of research universities, and compete in multiple countries simultaneously. This globalization of industry has occurred even as private-sector R&D in the United States as a percentage of GDP has increased and federal R&D has declined. As a result, other countries have much greater access to the results of the R&D done by U.S.-based companies than they did in the past.

THE EXPANSION OF CONTROL MECHANISMS

In response to growing competitive pressures from other countries, and in light of the U.S. policy framework emphasizing the role of restrictions on the *outputs* of the nation's technology enterprise, the U.S. research community has seen a steady increase in the number and complexity of restrictions governing the conduct of scientific and technology R&D. The simultaneous expansion of the number and scope of "critical technologies" raises important questions about whether restrictions alone are addressing challenges to U.S. technology leadership or if they are having increasing adverse consequences, since the United States is no longer dominant compared with its competitors.

Restrictions on Research Environments

Despite the uniform designation of information deemed to require protection as "controlled unclassified information" (CUI), the practices used by departments, agencies, and program managers continue to vary and consume large amounts of researchers' time. Researchers also are concerned about the imposition of controls in formerly uncontrolled environments.

Increasingly Complex Classification

Program managers at federal agencies often place clauses in research contracts designed to ensure that work that might require protection is properly identified prior to publication and to restrict the participation of foreign nationals in such programs. Such restrictions are clearly at odds with the intent of NSDD-189 to allow the free exchange of ideas. By limiting the exchange of ideas, participation by others, and international collaboration, such restrictions can slow the pace of research and make research environments less attractive to talented people. Unfortunately, the need to take precautions creates incentives for federal managers to impose restrictions but not for them to designate clearly work that can be done in unrestricted environments. It also has a chilling effect on researchers, who may be disinclined to work in certain areas or to speak on particular topics.

Increases in Security Requirements

The designation of information as CUI can have important implications for the conduct of research. To cite just one example, the federal government can handle CUI on information systems only if certain security measures are in place, and it can require nonfederal parties handling CUI to adhere to the same limitation. Many universities lack robust cybersecurity systems that meet these standards. Thus, requiring them to meet the standards for CUI can impose a considerable constraint on their research.

Restrictions on Foreign Collaborations

Congress and the Trump administration have taken several steps aimed at tightening security within the research enterprise (Goodrich, 2020a). For example, the Trump administration issued a proclamation limiting the entry of foreign researchers into the United States, imposing further restrictions in response to the COVID-19 pandemic (President of the United States, 2020). It specifically targeted Chinese graduate and postgraduate students and researchers with any ties to the Chinese government's "military–civil fusion strategy" (described in the next chapter).

Restrictions on Who Can Participate in Research

The above and other policy initiatives have placed a significant burden on U.S. research institutions and researchers. Complying with heightened rules and regulations reduces research productivity, with uncertain security benefits.

Focusing attention specifically on students and researchers from China has created mistrust and disruption that have further slowed research projects. Singling out researchers from particular countries has the effect of encouraging them to take their skills elsewhere. Given the globalization of science and technology, researchers may be able to do essentially the same work in those other countries, resulting in a net loss to the United States.

Increased Reporting Obligations

Funding agencies have increasingly placed more complete and detailed disclosure requirements on researchers, including reporting of all sources of support, conflicts of interest, and conflicts of commitment, and systematized those requirements across government.

In January 2022, a subcommittee of the Joint Committee on the Research Environment under the National Science and Technology Council released its "Guidance for Implementing National Security Presidential Memorandum 33 (NSPM-33) on National Security Strategy for United States Government-Supported Research and Development" (NSTC, 2022). The guidance was based on three principles: to protect America's security and openness, to be clear so that well-intentioned researchers can easily and properly comply, and to ensure that policies do not fuel xenophobia or prejudice. The guidance provides direction on five major areas of research security: disclosure requirements and standardization, the use of digital persistent identifiers, consequences for violations of disclosure requirements, sharing of information about research security, and research security programs at federally funded institutions. The guidance authorizes a requirement for federal agencies to implement NSPM-33 in a nondiscriminatory manner, including among members of ethnic or racial minority groups. It also notes that "agencies should incorporate measures that are risk-based, in the sense that they provide meaningful contributions to addressing identified risks to research security and integrity and offer tangible benefit that justifies any accompanying cost or burden" (NSTC, 2022, p. 1). The Office of Science and Technology Policy is expected to release standardized requirements for establishing research security later in 2022.[9]

Export Control Expansion and Designation of Essential Technologies

The Export Control Reform Act of 2018 called for the Commerce Department to establish export controls on "emerging and foundational technologies that are essential to the national security of the United States" (Fergusson et al., 2021). In establishing these controls, the Commerce Department was directed to consider the foreign availability of technologies, the effects on

---

[9] Letter from Alondra Nelson dated March 1, 2022, to Heads of Member Agencies of the National Science and Technology Council; https://www.whitehouse.gov/wp-content/uploads/2022/03/03-2022-Coordination_RS_Letter.pdf (accessed June 12, 2022).

development of these technologies in the United States, and the effectiveness of limiting the proliferation of these technologies. Although the legislation does not define the term "foundational," it is generally held to encompass security-relevant or economically relevant technologies on which future technology development depends.

In November 2018, the Commerce Department listed 14 categories of emerging technology as essential for national security and sought comment on this list:

- biotechnology;
- artificial intelligence and machine learning;
- position, navigation, and timing;
- microprocessors;
- advanced computing;
- data analytics technology;
- quantum information and sensing;
- logistics;
- additive manufacturing;
- robotics;
- brain–computer interfaces;
- hypersonics;
- advanced materials; and
- advanced surveillance technologies.

Many of those submitting comments urged caution, arguing that constraining the ability to develop such technologies and sell them could end up harming rather than benefiting the United States. Two years later, the Commerce Department sought public comment on the definition of and criteria for foundational technologies without offering specific candidates for comment. Many respondents similarly urged caution and pointed to the potential costs of controls on such technologies.

In February 2022, the Fast Track Action Subcommittee on Critical and Emerging Technologies of the National Science and Technology Council released a "Critical and Emerging Technologies List Update," which identifies critical and emerging technologies (CETs) as a "subset of advanced technologies that are potentially significant to U.S. national security" (Fergusson et al., 2021). Citing the 2021 *Interim National Security Strategic Guidance*, it defines three national security objectives: protecting the security of the American people, expanding economic prosperity and opportunity, and realizing and defending democratic values. The technologies it identifies are:

- advanced computing,
- advanced engineering materials,
- advanced gas turbine engine technologies,

- advanced manufacturing,
- advanced and networked sensing and signature management,
- advanced nuclear energy technologies,
- artificial intelligence,
- autonomous systems and robotics,
- biotechnologies,
- communication and networking technologies,
- directed energy,
- financial technologies,
- human–machine interfaces,
- hypersonics,
- networked sensors and sensing,
- quantum information technologies,
- renewable energy generation and storage,
- semiconductors and microelectronics, and
- space technologies and systems.

In each of these areas, it further identifies "key subfields." For artificial intelligence, for example, these subfields are

- machine learning;
- deep learning;
- reinforcement learning;
- sensory perception and recognition;
- next-generation artificial intelligence;
- planning, reasoning, and decision making; and
- safe and/or secure artificial intelligence.

The report states that "departments and agencies may consult this CET list when developing, for example, initiatives to research and develop technologies that support national security missions, compete for international talent, and protect sensitive technology from misappropriation and misuse." Given the very broad reach of the items on this list, its widespread application is likely to heighten controls on research. And despite the fundamental research exclusion on export controls, previous National Academies reports have noted that the current regime is unnecessarily burdensome and counterproductive to national security objectives, and has impeded university research in a wide variety of areas (NRC, 2007 and NASEM, 2016).

Controls on Foreign Investment

The Foreign Investment Risk Review Modernization Act of 2018 extended CFIUS's authority to review transactions beyond those conferring foreign ownership to include investments that might provide access to nonpublic technical information or data relevant to national security. These changes have directed the attention of CFIUS to "critical technologies," "critical infrastructure," and foreign investments perceived as harmful to national security (Jackson, 2020).

This expansion of the CFIUS review authority has required agencies to assign more staff to the review process to understand the technologies under consideration, potential risks, and possible mitigation measures. In general, the legislation marks an expansion of oversight beyond individual investment decisions to combinations of transactions and their effects on the U.S. economy and national security. It also raises significant questions as to whether CFIUS or CFIUS-like processes can operate on the timescales necessary to ensure that research collaborations can keep pace with the speed of new science and technology exploration and innovation. This issue is particularly acute at university laboratories, where the annual calendar associated with new graduates, postdoctoral researchers, and faculty would be seriously impacted by approval processes that span months or years.

IMPLICATIONS OF THE NEW COMPETITIVE LANDSCAPE FOR U.S. POLICIES AND PROCEDURES

U.S. policies, programs, and procedures designed to protect U.S. technology advantages have been proving less effective as the nation's competitive leadership in related areas of science and technology has narrowed. Protective laws, policies, and programs are often inconsistent; existing programs and agency approaches are often uncoordinated; and public–private coordination is weak. Furthermore, as discussed in Chapter 4, the United States now has a near-peer competitor—China—that has demonstrated its ability and willingness to devote substantial resources, both money and people, to enhancing its global position. In addition, China has a different social and governmental system and different ideological and cultural baselines, which together pose unprecedented challenges to U.S. competitiveness.

The United States still needs to protect a subset of technologies whose loss could result in a decline in national or economic security. But efforts to control particular "critical technologies" no longer constitute a sufficient strategy. Countries around the world have enhanced their own R&D capacity and are now active contributors to the development of many civilian and military technologies. The United States can no longer make decisions from a position of sole economic or military superiority. The nation's competitors have most of the tools available in the United States, and China and India have domestic human resources that surpass those of the United States. Global supply chains deliver technologies and materials vital for both national security and economic competitiveness; for example, the Department of Defense is dependent upon China and other countries for very large cast and forged parts, which are used in weapon systems and ground combat vehicles, among other defense applications (DoD, 2022). The United States also has major noncooperating adversaries that control key natural resources—for example, Russian natural gas, Middle Eastern petroleum, or Chinese rare earth minerals—allowing them to exert power beyond their economic or military standing.

In this context, the United States must now make a much broader effort to protect vital U.S. advantages in all aspects of technology development, commercialization, production, and use. The current policy landscape appears to be mismatched to the dual threats to U.S. interests of a more complex technology landscape and growing international competition in broad areas of research, development, and commercialization. Expanded efforts to protect or control a growing list of technology areas through restrictions on research, including areas of fundamental research and discovery, are not addressing ongoing concerns about these threats to U.S. technology leadership (President of the United States, 2020); instead, they may be adversely impacting U.S. leadership in science and research. Heightened rhetoric about the risks posed by immigrants or foreign-born researchers is motivating a number of restrictions and controls, including growing numbers of counterintelligence cases involving U.S. researchers, new restrictions on conflict management and disclosure, and limitations on visas on foreign-born students and researchers (Goodrich, 2020b). If the restrictions and controls instituted to address such concerns constitute the sole approach to technology protection and are applied to a broad range of research activities, the unintended consequence may be to erode the first-mover advantage in the discovery and development of the latest technologies enjoyed historically by the United States.

New circumstances require a new approach to technology protection. Focusing solely on limiting participation and controlling the outputs of the U.S. innovation system will not address the threats posed by competition in other areas. Current protection mechanisms can block the public- and private-sector innovation pipelines that drive economic competitiveness. Overly strict protections also constrain the access to talent from around the world that has been a pivotal factor in the nation's rise to global prominence. In today's world, a strong offense, not just a strong defense, has become the crucial determinant of success. While the battlefield has changed since the Cold War, leadership in science, technology, and innovation remains the most important weapon in the current competitive environment.

The potent combination of increased competition in research and innovation and the global interdependence of technology has made the current U.S. approach to protecting technology outdated. Openness and trust foster innovation leadership; the challenge in a world of interdependence and competition is to maximize openness and trust in a risk-conscious fashion. As important as the protection of certain specific technologies is, today's landscape requires a strong focus on protecting the *advantages* of the U.S. innovation system—those that help bring the fruits of the U.S. research and innovation ecosystem to market, with benefits for U.S. economic and security leadership.

4

The Competitive Challenge Posed by China

The competitive challenge that China poses to the United States is unprecedented. Through its large investments in research and development (R&D) funding and personnel and its leaders' attention to science and technology, China has indicated that its ambitions are not just to catch up with the United States in science and technology but to surpass it. This chapter highlights aspects of the competition between the United States and China and the implications of China's actions for U.S. interests.

Many countries support public- and private-sector R&D to strengthen their economies and national security. The United States is still the largest single funder of R&D globally, but it now accounts for only about a quarter of the global total (Figure 4-1) (NSB, 2020). International collaboration, digital communications, and the flow of students and STEM (science, technology, engineering, and mathematics) professionals across borders augment countries' R&D expenditures by ensuring the rapid dissemination of new ideas and information. The United Nations' Sustainable Development Goal 9.5 calls on nations to "enhance scientific research, upgrade the technological capabilities of industrial sectors in all countries, in particular developing countries, including, by 2030, encouraging innovation and substantially increasing the number of research and development workers per 1 million people and public and private research and development spending." Scientific research and technology development are worldwide endeavors that have widespread benefits even as they improve the economic conditions and military strength of particular countries.

In this global science and technology enterprise, China has recently assumed a particularly prominent position. Funding for R&D in China has risen 30-fold since 1990 and, given a continuation of current trends, will surpass U.S. funding during the early part of this decade (Committee on New Models for U.S. Science and Technology Policy, 2020). Chinese scientists now publish more scientific articles than do U.S. scientists, although papers from U.S. scientists are currently more highly cited overall (NSB, 2020b).[1] Chinese universities have been producing more STEM Ph.D.'s than U.S. universities for more than a decade, and by 2025 are expected to be producing nearly twice as many (Zwetsloot et al., 2021). Compared with their U.S. counterparts, China's leaders devote much more attention to science and technology, and China's growing strengths in science and technology are a source of national pride. For all these reasons, China is an excellent example of the new kinds of competitive challenges facing the United States.

This chapter examines the steps China has taken and is planning to take to become a global leader in science and technology. Using synthetic biology and the other cases studied as examples (see Chapter 2), it looks not just at R&D funding but also at China's efforts to acquire technologies developed elsewhere, attract top talent, and establish its supremacy in both targeted areas of science and technology and in the overall strength of its science and technology system. Discussed as well are the large numbers of Chinese undergraduate and graduate students and postdoctoral fellows who come to the United States to study science and engineering and do scientific research, many of whom remain in the United States to work.

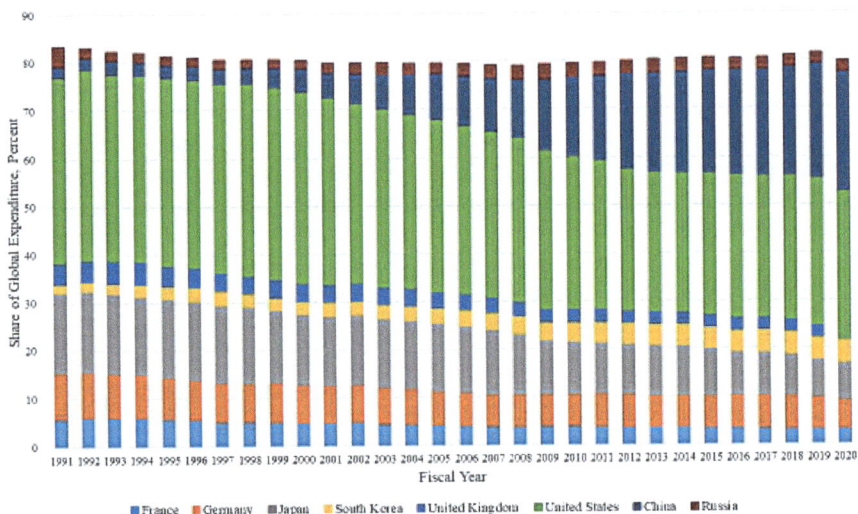

FIGURE 4-1 Share of global research and development funding, 1991–2020, by country.
SOURCE: Based on data from OECD, 2022.

---

[1] The National Science Board (NSB) notes that between 2000 and 2020, the index of highly cited articles (a country's share of the top most-cited science and engineering (S&E) publications divided by a country's share of all S&E publications) for the United States remained stable at 1.8, while China's index increased from 0.4 to 1.2. An index over 1.0 indicates that a country contributes a larger share of top-cited articles, compared with its share of overall publication output.

FEATURES OF THE COMPETITION BETWEEN THE UNITED STATES AND CHINA

The competition between the United States and China for leadership in technologies strategically important to national and economic security differs in many respects from the competition between the United States and Soviet Union during the Cold War. Unlike the Soviet Union, China has learned how to leverage domestic and foreign demand to enhance economic growth, and its economy has been expanding rapidly. China and the United States are much more intertwined economically than were the United States and the Soviet Union, with the economies of each relying heavily on exports and services provided by the other. Many Chinese students attend U.S. universities, and many scientists and engineers born and raised in China work in the U.S. science and technology enterprise, which was not the case with Soviet scientists during the Cold War. Each country plays outsized and interdependent roles in the major challenges facing the world today, such as climate change and global pandemics. Each country is part of broader economic and security networks that both increase capabilities and constrain actions.

The current U.S. competition with China is also different from the economic competition between the United States and Japan in the 1980s. Japan was competing in some of the economic sectors in which the United States had long held dominance, such as microelectronics, automobiles, and machine tools; however, Japan was a geopolitical ally with similar innovation, economic, and governance systems. Japan's innovation system was also smaller, with Japanese graduate students and postdoctoral fellows having a small presence in U.S. universities compared with the much larger numbers of Chinese students in U.S. universities today. While the Japanese government played an active role in Japan's competitive position, including subsidizing its domestic industries and seeking to acquire technology from abroad, it adhered to international norms on economic, legal, and trade matters.

Although the United States and China are the two largest funders of R&D in the world today, aspects of the two countries' approaches to supporting science and technology differ substantially. First, as discussed further below, in its effort to reduce its reliance on foreign technologies and assume leadership in the technologies of the future, China engages in much more integrated planning and strategic action relative to the United States. The Made in China 2025 technical area roadmap, for example, calls for China to become a leader in ten industrial sectors (Alves Dias et al., 2019):

- next-generation information technology,
- high-end numerical control machinery and robotics,
- aerospace and aviation equipment,
- maritime engineering equipment and high-tech maritime vessel manufacturing,

- advanced rail equipment,
- energy-saving vehicles and new-energy vehicles,
- electrical equipment,
- agricultural machinery and equipment,
- new materials, and
- biopharmaceutical and high-performance medical devices.

Besides investing heavily in R&D to develop commercially and militarily valuable technologies, China has been seeking to acquire technologies developed elsewhere (Brown and Singh, 2018). Chinese companies have purchased technology companies in the United States and elsewhere to gain access to cutting-edge research. The Chinese government has subsidized domestic industries, has blocked foreign investment in Chinese companies to give those companies an advantage in foreign competition, and engages in industrial espionage and cybertheft to acquire technology. When foreign investment in Chinese companies is allowed, technology transfer is often a condition of building, owning, or operating facilities in China. China has been acquiring more power in the United Nations and other international bodies, including standards-setting bodies. It has an integrated and top-down set of strategies, and it takes a whole-of-government and global approach to implementing those strategies, including planning for raw materials and supply chain issues while a technology is only in the research phase.

Another difference between the United States and China is China's explicit advocacy of what it calls "military–civil fusion," which is aimed at eliminating barriers between its commercial and military sectors (Kania and Laska, 2021). Although the implementation of this strategy faces significant obstacles, its goal is to integrate economic development and military modernization by having the commercial and defense enterprises share innovations, resources, and talent. China's commitment to the concept of fusion signals its desire to achieve integrated commercial and military leadership. The United States does not have an official policy to promote interactions between the civilian and defense sectors, although many companies work in both sectors, companies in one sector interact with companies working in the other, and many university research labs conduct defense-related research funded by the government. In addition to federal laboratories associated with the Department of Defense, the United States also maintains 17 national laboratories housed in the Department of Energy and staffed by civilian contractors, and many of these laboratories play a significant role in support of the defense mission (DOE, 2020).

The Chinese government has been emphasizing the ways in which it is distinct from the United States and other democracies. At the time China entered the World Trade Organization, many expected it to become a more open and market-oriented country (Mavroidis and Sapir, 2021); instead, it has taken a turn toward authoritarianism and military assertiveness (Liff and Ikenberry, 2014). While the United States and other nations are committed to individual freedoms and to a diversity of identities, China is committed to the belief that all its citizens should share a fundamental identity and allegiance to the state. It has sought to draw other countries, particularly those with authoritarian leaders, into closer ties by emphasizing its success in lifting hundreds of millions of people out of poverty. China has sought to portray democratic countries as failing experiments in governance while touting and advancing its own achievements and worldview.

As noted previously, a distinguishing feature of China's approach to gaining market leadership over the West has been its use of detailed strategies outlining the specific technology areas in which it seeks to attain leadership; outlined as well are the steps it is taking to that end. This approach is quite different from that of the United States, which has historically based its actions on broad precepts, such as the expansion of democratic principles and expansion of free trade. U.S. strategies have historically not been tied to specific technology areas, with the exception of narrow areas in which technologies have clear military or national security applications, as is the case with, for example, nuclear weapons technologies.

China's advantages in research, development, and innovation do have limitations. China's government, economy, and society have many weaknesses, including inefficient state-owned enterprises; a lopsided demographic structure, with 120 boys born for every 100 girls; rising levels of debt; strong neighboring countries; a dependence on imported goods; and a rigid ideology that limits experimentation (Hass, 2020). China's reliance on other countries for raw materials, markets, and advanced training in science and technology limits its ability to act unilaterally. The U.S. science and technology ecosystem is still stronger and more agile than China's, in part because of diversified federal support for R&D, the strength of U.S. capital markets, and human resources from domestic and international sources.

SYNTHETIC BIOLOGY IN CHINA[2]

A closer look at China's initiatives in one of the four case studies examined in Chapter 2—synthetic biology—provides a specific example of the steps that country is taking to attain leadership in science and technology. A high-level Communist Party and Chinese Academy of Sciences official voiced China's intentions, stating, "As Europe won in the 19th century using industry, and the United States won in the 20th century using information technology, so China will win in the 21st using biology" (Carlson, 2019).

The United States is the global leader in synthetic biology, largely as a result of massive investments in biology and life sciences research and especially since the doubling of funding for the National Institutes of Health (NIH) in the

---

[2] This section is based in part on the presentations at a workshop on synthetic biology held by the committee on May 13, 2021. An agenda for the workshop and speaker biographies are available at https://www.nationalacademies.org/event/05-13-2021/protecting-critical-technologies-for-national-security-in-an-era-of-openness-and-competition-meeting-3-workshop-on-synthetic-biology.

early to mid-1990s (NASEM, 2020b). As discussed in Chapter 2, the significance of global leadership in this area goes beyond the development of biologic materials and products: synthetic biology is best viewed as a broad "production platform" with the potential to disrupt a wide range of technology areas (El Karoui et al., 2019).

Synthetic biology–related research in China started as early as the 1960s, when Chinese scientists produced synthetic insulin (Kung et al., 1965). Despite some early ventures, however, serious interest in synthetic biology did not begin in China until 2007, when a Chinese team became world champions at an International Genetically Engineered Machines (iGEM) competition with their project "Towards Self-Differentiated Bacterial Assembly Line" (Moshasha, 2016). A year later, China held the Xiangshan Conference on Synthetic Biology, during which Chinese biologists emphasized the massive contributions that synthetic biology could make to the future bioeconomy and called for greater government support for the field.

Following this conference, the Key Laboratory of Synthetic Biology was launched in 2008, marking the Chinese government's first official foray into synthetic biology. Established by the Chinese Academy of Sciences, the laboratory sought to design functional biological parts that could produce biomaterials and bioenergy through the modification and synthesis of biological systems. Since then, synthetic biology has advanced at a rapid pace in China, with an early focus both on bioremediation to improve the lives of Chinese citizens and on the growth of the Chinese bioeconomy. By 2013, when the U.S. National Academy of Engineering and National Research Council held a series of symposia on synthetic biology, China was already contributing about 10 percent of papers on synthetic biology published globally each year (NAE and NRC, 2013).

Today, numerous Chinese organizations, largely government-related—including the Chinese Academy of Sciences, the Chinese Academy of Engineering, the national and local offices of the China Academy of Machinery Science and Technology, and medical universities—support research in synthetic biology. Funding for this research comes from many sources, with estimates totaling roughly U.S. $100 billion per year (NAE and NRC, 2013).

In 2009, the Chinese Academy of Sciences developed a strategic roadmap, "Innovation 2050: Technology Renovation and the Future of China," outlining desired achievements in technology, industrial applications, medicine, and agriculture within 5, 10, and 20 years (Pei et al., 2011). Part of this roadmap focused on synthetic biology, including "goals related to the availability of comprehensive databases for synthetic parts, a timeframe for commercial application of engineered parts, and a timeframe for clinical application of devices and systems" (NAE and NRC, 2013, p. 18). China's 13th Five-Year Plan established a goal for biotechnology to contribute 4 percent of the country's gross domestic product (GDP) (or roughly U.S. $600 billion) by 2020 (*People's Daily*, 2017*)*.

Biotechnology is also becoming increasingly important in Chinese military doctrine, with the People's Liberation Army designating biology as a separate warfighting domain (Cunningham and Geis, 2020). Potential applications to the Chinese military include biomaterials, human enhancement, and offensive capabilities that may include ethnically targeted bioweapons (Cunningham and Geis, 2020). Thus far, the United States has not focused on the potential for biotechnology to transform offensive military technology, thereby creating an opportunity for China to gain a military advantage.

The Chinese government is seeking access to foreign capabilities in synthetic biology to accelerate the development of its domestic industries, in part through the acquisition of foreign companies and technologies (FBI, 2019; see also Brown and Singh, 2018; Sganga, 2022). It also is developing resources that it is carefully protecting from other countries. For example, China has collected the world's largest human genetic database and has prohibited its export to preserve its intrinsic economic and security value (Cunningham and Geis, 2020). In general, as noted earlier, China wants to ensure that it does not just catch up to the United States technologically but surpasses it to dominate the technical field (Huggett, 2019).

In the past decade, the Chinese government has supported the development of numerous academic centers for synthetic biology. In 2017, the Chinese Academy of Sciences launched the Institute of Synthetic Biology, the country's first institute in this field. Guided by the principle "build life for understanding it, build life for applications," the institute was designed to integrate research in biotechnology and information technology to further understanding and applications of synthetic biology. Today, the institute is home to the world's largest cutting-edge multidisciplinary team in synthetic biology, composed primarily of young "overseas returnees" specializing in the field.

In 2018, the Shenzhen Institute of Advanced Integration Technology (SIAT) approved the combined development of the Shenzhen Institute of Synthetic Biology and the Institute of Synthetic Biology at SIAT. With a 750 million RMB (about U.S. $110 million) investment from the Shenzhen Municipal Government, SIAT quickly built one of the world's largest scientific and technological facilities for research in synthetic biology. This facility houses numerous synthetic biology research centers, including the Center for Quantitative Synthetic Biology, the Center for Synthetic Genomics, the Center for Synthetic Biochemistry, the Center for Synthetic Microbiome, the Center for Genome Engineering and Therapy, the Center for Synthetic Immunology, the Materials Synthetic Biology Center, and the Center for Cell and Gene Circuit Design. The main building, completed in January 2021, serves as an advanced platform for the design and fabrication of biological systems (Shenzhen Institute of Synthetic Biology, n.d.).

In 2019, backed by billionaire Li Ka Shing's HK $500 million (U.S. $70–80 million) donation, the Hong Kong University of Science and Technology (HKUST) launched the Li Ka Shing Institute for Synthetic Biology. The institute, which emphasizes originality and the application of foundational knowledge, aims to integrate genetic engineering with artificial intelligence (AI) and relevant analysis methodologies to bring about discoveries that lead to innovative products

(Cumbers, 2019). HKUST expects to invest about U.S. $1 billion in the institute during this decade to make Hong Kong a global hub for synthetic biology. This investment aligns with the Chinese government's strategy to go from serving as the world's factory to developing advanced-value products, thereby advancing domestic resilience, improving and stabilizing industrial supply chains, improving advanced manufacturing and promoting scientific research, coupling basic life sciences research with information technology, ensuring harmony between humans and nature, and strengthening the country's public health system (Tsang and Poon, 2021). The United States has synthetic biology centers housed at the Massachusetts Institute of Technology; the University of California, Berkeley; Northwestern University; and elsewhere. Nonetheless, the gift from the Li Ka Shing Foundation to HKUST represents the largest single donation for synthetic biology research globally, reinforcing China's belief in the importance of the field to its future global economy.

In the past decade, China has also made significant investments in bio databanks and biofoundries. Established in 2011, the China National GeneBank (CNGB) was China's first national-level gene storage bank, approved and funded by the Chinese government. The Center for CNGB, including a biorepository, a bioinformatics data center, and a living biobank, opened in Shenzhen in 2016. Combining these repositories, the China National GeneBank DataBase (CNGBdb) was designed to provide a unified platform for biological big data sharing and application services to the research community.[3] Using big data and cloud computing technologies, CNGBdb has integrated large amounts of internal and external molecular information, and has correlated living sources, biological samples, and bioinformatics data to enable biological data to be traced throughout the life cycle. In collaboration with Australia's Macquarie University and Harvard University, CNGB also has a synthetic biology platform focused on metabolic engineering and the development of high-density DNA storage technology (GenomeWeb, 2018). Today, China has accumulated the largest genomic holdings of any country in the world (Ratnam, 2021).

Founded in 2019, the National Genomics Data Center (NGDC), part of the China National Center for Bioinformation (CNCB), was designed to provide open access to a suite of data resources and services generated from large-scale sequencing studies on precision medicine and biodiversity. Since the beginning of the COVID-19 pandemic, for example, the CNCB-NGDC has focused in particular on building a SARS-CoV-2 information resource through genomic data collection, curation, and deep mining with extensive daily updates. This database, named the 2019 Novel Coronavirus Resource, contains an open-access, comprehensive collection of genome sequences and clinical information for all publicly available SARS-CoV-2 isolates. Other new databases that have emerged from the NGDC include the Aging Atlas, an integrative database designed to support research on aging; Brainbase, a curated knowledge base for brain

---

[3] China National GeneBank DataBase (https://doi.org/10.25504/FAIRsharing.9btRvC [accessed June 10, 2022]).

diseases; and the Chloroplast Genome Information Resource, a curated resource of chloroplast genome information. Simultaneously, a number of resources have been updated and improved, including BioProject, BioSample, and several biodiversity and plant resources. Of particular note, BIG Search, a scalable, one-stop, cross-database search engine, has been updated to provide easy access to a large quantity of internal and external biological resources (CNCB-NDGC Members and Partners, 2020).

In 2020, Shenzhen Science City announced that the Advanced Biofoundry Shenzhen will be one of the key projects for priority launch. This biofoundry will have three platforms: a design–learn platform, a synthetic testing platform, and a user testing platform (Shenzhen Institute of Advanced Technology, 2020). Ultimately, the Shenzhen biofoundry is intended to expedite the "design–build–test–learn" cycle economically to realize the rational design and synthesis of artificial living systems.

With these and other recent developments, China has created and is pursuing a coherent national plan for challenging the U.S. leadership in the synthetic biology field. Currently, the United States and China remain interdependent and closely intertwined in the field: the United States relies on China for manufacturing, services, and talented students who come to study and work at U.S. universities, while China depends on external basic research to support a bioeconomy geared toward commercialization of innovations created elsewhere. Decoupling the activities of the two countries would be difficult, but China is clearly striving for dominance in synthetic biology. If the United States is to remain competitive in synthetic biology, it will need to remain vigilant about China's activities in the field and develop a comprehensive strategy for responding to the competitive challenge posed by those activities.

CHINA'S ACTIVITIES IN MICROELECTRONICS, ARTIFICIAL INTELLIGENCE, AND QUANTUM COMPUTING

Less in-depth examinations of the other three case studies considered in Chapter 2 similarly reveal the efforts China is making to gain leadership in microelectronics, AI, and quantum computing. In contrast to China's directed and coordinated planning to achieve positions of technological leadership, the United States generally has not developed a strategic policy for advancing innovation and commercialization of these technologies.

Microelectronics

Although China is not yet among the world's leaders in the design or production of leading-edge microelectronics, it is narrowing the gap with the leaders (Graham et al., 2021). China's semiconductor manufacturing capacity has already surpassed that of the United States, and China is projected to become the world's largest semiconductor manufacturer by 2030. It also is the largest single consumer of semiconductors, creating powerful incentives for it to advance its domestic microelectronics industry.[4]

According to a recent report by the U.S. Trade Representatives, "China's strategy calls for creating a closed-loop semiconductor manufacturing ecosystem with self-sufficiency at every stage of the manufacturing process—from IC [integrated circuit] design and manufacturing to packaging and testing, and the production of related materials and equipment" (USTR, 2018, p. 113). The national and provincial governments are supporting "national champion" firms under a National Integrated Circuit Investment Fund to acquire critical technologies and build advanced fabrication facilities (Kim and VerWey, 2019). The Made in China 2025 technical area roadmap calls for the main segments of the industry to reach advanced international levels by 2030. At that point, domestic producers should be providing 80 percent of domestic consumption of integrated circuits.

The Chinese government directly supports the industry through favorable loans, direct grants, reduced utility rates, tax breaks, and free or discounted land (SIA, 2021). These incentives have spurred the creation of thousands of new semiconductor companies, and more than 100 new fabrication facility projects have been announced since 2014. China is also supporting other parts of the microelectronic supply chain in an effort to achieve indigenous capabilities. As the Semiconductor Industry Association has stated, "If left unchecked, state-owned Chinese firms shielded from market forces, or [with] access to illicitly acquired IP [intellectual property], could pose significant challenges to the health of the U.S. semiconductor industrial base" (SIA, 2021, p. 6).

Artificial Intelligence

The New Generation Artificial Intelligence Development Plan (AIDP) released by the Chinese government in 2017 called on China to "plan, grasp the direction, seize the opportunity, lead the world in new trends in the development of AI, serve economic and social development, and support national security, promoting the overall elevation of the nation's competitiveness" in AI (China State Council, 2017). By 2025, according to the plan, "China will achieve major breakthroughs in basic theories for AI, such that some technologies and applications achieve a world-leading level and AI becomes the main driving force for China's industrial upgrading and economic transformation" (China State Council, 2017). Under the plan, China will become the world's innovation center for AI by 2030.

With active support from the highest levels of government, China is pursuing a variety of AI applications in both the military and commercial sectors (Allen, 2019). It has established two major new research organizations on AI and

---

[4] Although China assembles more than one-third of the world's electronic devices, its share of electronic device end users is nearly as large as that of the United States (SIA, 2021).

unmanned systems under the National University of Defense Technology, and private-sector companies are developing competitive products and services that incorporate AI technologies. Chinese AI researchers are involved in many international collaborations, with more than half of Chinese AI papers being coauthored with non-Chinese authors. To develop the country's human resources in AI, the Chinese Ministry of Education has developed AI research centers, open courses, and teaching materials. Drawing on China's experiences with telecommunications, Chinese companies and government organizations are working to shape standards on AI to support economic growth and national security.

China has certain advantages over other countries that are individually developing AI, such as weak data protection regulations that have allowed for the collection and sharing of large amounts of personal data (Huw et al., 2021). At the same time, the AIDP calls for China to become a leader in the development of ethical norms and standards for AI, including respect for human rights, privacy, and fairness, although efforts to do so are in the early stages.

Quantum Computing

In 2016, China launched a "megaproject" aimed at making breakthroughs in quantum computing; its funding for quantum information sciences now greatly exceeds that of the United States (Graham et al., 2021). In the country's 14th Five-Year Plan, both quantum computing and AI are cited as high priorities (CSET, 2021). Although there are different methods of measuring quantum computing capabilities, many individuals using open-source information believe that the capabilities of quantum computers in China now rival the capabilities of those in the United States, and Chinese researchers and organizations now have more patents in the field than do Americans (Graham et al., 2021). And many agree that China has already taken the lead in quantum communications (Kwon, 2020). China has announced that it intends to surpass the United States in quantum technologies and their applications in the military and commercial sectors (CSET, 2021).

In 2021, the U.S. Commerce Department, worried about the encryption of sensitive U.S. communications, blocked U.S. firms from exporting quantum computing technology to eight Chinese companies and laboratories. However, China's stated desire to lead the world in quantum technologies and its cultivation of talent in the field indicate that technology controls may do little to slow its advance (Kania and Costello, 2018).

HUMAN RESOURCES IN THE UNITED STATES AND CHINA

The productivity and quality of the scientific research and technological development carried out in a country reflect the education, creativity, and dedication of the people doing that work. In this respect, the United States has been singularly fortunate. Bolstered by the nation's historical strengths in manufacturing, engineering, industrial organization, and innovation, STEM fields have historically attracted many talented U.S. students for both study and work. Over the course of the 19th and 20th centuries, scientists and engineers educated in the United States transformed the nation from an afterthought in the global scientific enterprise to a science and technology powerhouse.

In addition to the nation's domestic sources of talent, many gifted and accomplished scientists, engineers, and entrepreneurs from other countries have come to the United States to study, do research, teach, work in industry and government, start companies, and otherwise contribute to the U.S. economy and society. Since the first Nobel prizes were awarded in 1901, Americans have accounted for more than a third of the approximately 900 recipients of the prize—nearly three times as many as the second-place nation, the United Kingdom—and about a third of the prizes in physics, chemistry, and medicine have gone to scientists who immigrated to the United States (NFAP, 2019). While Nobel prizes are just one measure of a nation's competitiveness, the U.S. dominance in this regard reflects the country's historical attractiveness as a place to learn and do science.

International Students in STEM Fields

As noted in Chapter 3, the number of international graduate students in U.S. colleges and universities has grown rapidly in recent decades, rising from under 250,000 in 2000/2001 to close to 400,000 in 2016/2017, although immigration policies and COVID-19–related restrictions have coincided with a slight decline in those numbers in recent years (Figure 4-2) (Israel and Batalova, 2021; Open Doors, n.d.). The number of international students enrolled in U.S. universities declined in 2020 as a result of the COVID-19 pandemic, but it declined less in graduate-level science and engineering than in other disciplines. Students on temporary visas earn about one-third of the doctoral degrees in science and engineering awarded by U.S. universities, and they earn more than half in some fields, including engineering and computer science (NSB, 2022b).

In the 2018/2019 academic year, an estimated 16 percent of U.S. graduate students in STEM fields—about 40,000 students at the master's level and 36,000 Ph.D. students—were Chinese (Feldgoise and Zwetsloot, 2020). These 76,000 students represented about 37 percent of all the international graduate students at U.S. universities. The number of Chinese graduate students more than doubled in the decade before 2018/2019, although the rate of growth slowed toward the end of that period.

In surveys of international Ph.D. recipients, more than 70 percent of those in STEM fields have expressed their intention to stay in the United States to work, with the highest rates (more than 85 percent) in computer science, biology, and engineering and among students from China, India, and Iran (Zwetsloot et al., 2020). Intention-to-stay rates, which correlate closely with actual stay rates, either remained steady or increased slightly between 2000 and 2017. Among recent cohorts of international Ph.D. recipients who had temporary

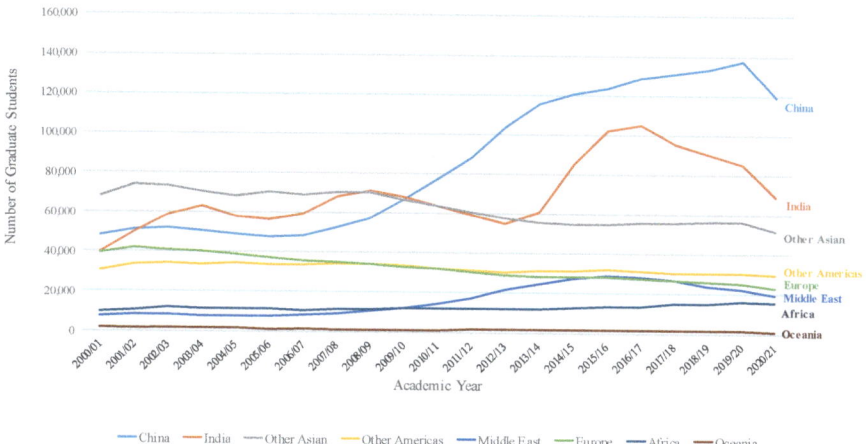

FIGURE 4-2 Number of international graduate students in the United States, 2000–2021, by place of origin.
SOURCE: Based on data from the Institute of International Education (2021).

visas at graduation, 70 percent were still in the United States 5 years later, and 62 percent were still here 10 years later (Finn and Pennington, 2018). Foreign-born workers remain an important part of the U.S. STEM workforce, representing 20 percent of that overall workforce and a much higher percentage in some fields.

In recent years, international student enrollment in such nations as Australia and Canada has increased as much as 20 percent, while that in U.S. institutions has flattened. This pattern appears to reflect a mismatch between immigration rules and talent needs in the United States, with inadequate attention being paid to matching employer needs and work/residency visas. In contrast to the immigration systems of other English-speaking countries, such as the United Kingdom, Australia, and Canada, the system in the United States is less focused on skills-based needs (NASEM, 2015). Permanent resident status in the United States is granted mainly to immigrants who are sponsored by family members instead of being employment based (Cohn and Ruiz, 2017). According to the most recent data released by the Department of Homeland Security, only 12 percent of new permanent residents were in management, professional, and related occupations, while 26 percent were not working outside the home (DHS, 2019). The number of highly skilled immigrant workers is growing more rapidly in other Organisation for Economic Co-operation and Development (OECD) countries than in the United States, and the U.S. share of international students has been declining. Undergraduate and graduate students continue to come to the United States, but many of these students study on temporary visas and must return to their home countries 6–9 months after receiving their degrees. In addition, a long tradition of postdoctoral fellows in scientific disciplines coming from Europe to train in the United States has eroded in recent years, in part as a result of the shifting political landscape (Israel and Batalova, 2021). Recent immigration reforms, such as the expansion of optional practical training, have allowed foreign students to work longer in the United States after completing their degrees, but the number of temporary work visas is capped, even in fields in which workers are in high demand.

Other countries, particularly China, have created incentives to attract students educated and researchers working in the United States to return to their home countries (Permanent Subcommittee on Investigations, 2019). China has invested billions of dollars in research funding, laboratory space, salaries, and other incentives to recruit scientists working abroad. It also has greatly increased its investments in higher education in an effort to create world-class universities in China so that it will have to rely less on other countries for advanced training.

When researchers educated or working in the United States move to China, they take information, experience, and know-how with them—an inevitable consequence of international flows of talent. The movement of people and knowledge from one country to another links the R&D enterprises of nations while also benefiting the recipients of talent. When students return to their home countries, they tend to retain contacts with colleagues from the countries they left, and many draw on these networks in doing research and starting businesses (Saxenian, 2006). At the same time, the home countries are more likely to benefit from the initiative and insights of scientists, engineers, and entrepreneurs returning home from abroad.

International Collaboration

Another way in which China has sought to strengthen its science and technology enterprise is by forming partnerships with researchers working in other countries. For example, China's Thousand Talents Program, launched in 2008, has recruited more than 7,000 native Chinese and foreign-born scientists to work with researchers in China and spend at least part of every year in that country. Now known as the National High-end Foreign Experts Recruitment Plan, the program seeks to attract high-level researchers to permanent or temporary appointments in China. The U.S. government has expressed concern about the program's being used to transfer know-how and intellectual property to China and has prosecuted some U.S. scientists for failing to disclose connections to the program.

In 2018 the Trump administration announced a new program, which came to be called the China Initiative, aimed at preventing economic espionage (Aloe and Guo, 2022). The Biden administration has since redirected the program to one aimed more broadly at threats from hostile countries, but the program still retains a focus on academic researchers with ties to China working in the United States. The program has been heavily criticized for charging researchers with lesser infractions than espionage and for ethnic profiling (Mervis, 2022). As described in Chapter 3, the guidelines released by the National Science and

Technology Council in January 2022 were designed to clarify the kinds of international relationships that are allowed and the disclosures that researchers must make about those relationships, although these guidelines are still in the process of being implemented (NSTC, 2022).

In response to concerns about illicit transfers of know-how and technology to China and other countries, universities have reemphasized and strengthened policies and procedures to address security threats and undue influence of foreign government on campus. They have increased training of faculty members and students, heightened protections for data and intellectual property, reviewed collaborations and contracts, and enforced conflict-of-interest and foreign-travel policies, among other measures.

The U.S. federal laboratory system also attracts some of the top scientific and technical talent both nationally and globally, and develops and maintains unique scientific resources not found elsewhere, from high-performance computers to specialized fabrication facilities. Team science is performed on a scale that is difficult to support in traditional universities or industry. Both open and deeply classified work is performed, and there are opportunities for foreign nationals as well as U.S. citizens. Investments in these institutions is part of how the United States cultivates talent and innovation (NASEM, 2021).

The Human Resources Challenge for the United States

To maintain its leadership in science and technology, the United States will need to cultivate both domestic and international talent. Domestically, the United States needs to invest in American STEM education programs to take greater advantage of the talent inherent in U.S. students. Scholarships, internships, targeted hiring practices, postdoctoral fellowships, and early-career opportunities can all increase the participation of U.S. students in STEM programs. Young people are passionate about solving real-world problems, and science and technology offer them ways to work on such problems. However, attracting the best and the brightest to these fields requires appropriate career and financial incentives. Students need to be recognized and rewarded for their efforts and achievements, and they need to have the promise of a good life if they are to pursue STEM careers. For example, the very low salaries paid to postdoctoral fellows in some fields are impeding the development of the U.S. science and technology workforce. The United States also needs more workers with other STEM skills, such as technicians and laboratory safety specialists. Increased government support for STEM partnerships between community colleges and industry could help grow the technically skilled workforce. Such collaborations could also create employment opportunities in regions where traditional employment opportunities have lagged, creating new opportunities for communities to grow and prosper.

The United States will not be able to rely entirely on domestic students to remain a leader in science and technology; it will need to continue to attract and retain students from around the world to realize the massive benefits provided in the past by talent from abroad. The United States has many advantages—including its open and thriving research institutions, its risk-taking environment, and the personal and professional rewards available for achievement—that appeal to smart and ambitious students. But researchers from abroad would be more likely to stay in the United States if they could bring their families, which would require changes in visa processes. Clearance processes could begin early so that researchers from other countries would be ready to do classified work if the need arose. Today, international students come to U.S. colleges and universities under the condition that they agree to return to their home countries upon graduation. This policy is at odds with the need to retain talent and expertise in the United States.

IMPLICATIONS OF CHINA'S ACTIONS FOR THE PROTECTION OF U.S. INTERESTS

The competitive challenge China poses to the United States is unprecedented. China's large investments in R&D and higher education, its talent programs, its scientific and technical intelligence efforts, and its leaders' attention to science and technology all demonstrate the country's ambitions not just to catch up with the United States in science and technology but to surpass it.

To safeguard, enable, and strengthen its national advantages in science and technology, the United States must take coordinated and comprehensive actions. For both economic and military reasons, the United States has a relatively small number of technologies that it wants to protect rigorously from being acquired by China and other countries. However, broadly applying mechanisms to protect technologies is counterproductive in that doing so impedes U.S. advances in science and technology more than it blocks the international diffusion of technologies.

The most important actions the United States can take are ones that will bolster its own scientific and technological competitiveness. These actions might include an integrated strategy for maintaining U.S. competitiveness that incorporates stronger institutions for technology development and application, enhanced efforts to attract the best students from around the world to U.S. universities (including a compelling narrative that challenges the Chinese narrative of a United States in decline), and increased R&D funding. In addition, any U.S. strategy will need to respond to explicit Chinese strategies that are technology specific and outline direct steps that China is taking to overtake the United States, an effort that will need to start with better monitoring of the actions taken by China and other countries (Brown and Singh, 2018).

5

Findings

As discussed in the previous chapters, international challenges facing the United States differ from previous national and economic security risks in a number of ways:

- Today's technologies are increasingly based on broad and enabling platforms, whereby new applications, specific technologies, or processes are built on shared and reusable elements that provide enormous advantages in speed and/or scale for discovery, development, supply chains, and/or production.
- Those platforms are increasingly being developed in the private sector instead of being initially developed by the government.
- The nature of global competition has changed. The United States faces an adversarial near-peer competitor in China, with investments in research and development (R&D) on par with U.S. investments, a well-educated labor force triple the size of that of the United States, and a worldview different from that of the United States and its allies.
- Scientific research and technology development have become much more internationally distributed and integrated.
- Increases in R&D spending and STEM (science, technology, engineering, and mathematics) education in other countries have expanded the amount of research being done in academic settings outside the United States, along with an almost instantaneous flow of information across borders and within the United States.
- Industrial research and production have become globalized, because firms either have become multinational enterprises with affiliates and customers in many countries or are increasing offshore research and production.

The above differences, coupled with the increasing speed of both information flow and technology development and application, create challenges for the United States in both developing and maintaining the strength of its technology ecosystem and controlling the loss of technologies and information via both legal and illegal pathways. The United States will need to find solutions in *policy; structure; focus; investment in research, infrastructure, and human capital;* and *governance*—including solutions in both the public and private sectors—if it is to retain the technological advantages that have historically underpinned both national security and a vibrant, open, and agile innovation engine. In a world where change is increasingly driven by platform technologies, the speed with which new capabilities can be applied in both the national security space and the competitive commercial marketplace and the shorter life cycles of technical capabilities create a very different timeframe within which competitive technical dominance can be maintained. Thus, *policies, structures, foci, investments*, and *governance* tools must emphasize the removal of impediments to innovation and speed.

The analyses and case studies presented in Chapters 2 through 4 lead to 13 key findings.

Finding 1: The historical approach to risk and protection of technologies taken by the United States has been predicated on an assumption of economic and technological dominance, largely by focusing those approaches on restricting access to or use of the outputs of technology development. The increasing globalization of science and technology, new ways of developing technology applications, and the advent of powerful technology platforms have made many of the current methods of protecting technologies obsolete, and in many instances counterproductive.

Stemming from its world-leading R&D enterprise, its attraction of top talent from around the world to work in that enterprise, and a culture of risk taking and entrepreneurship that has fostered the translation of ideas into new products or services, the United States has enjoyed long-standing dominance in the capability to develop and commercialize new technologies. As discussed in Chapter 3, the United States emerged as the world's dominant economic power after World War II, accounting for 40 percent of the world's gross domestic product (GDP) in the 1960s while having less than 6 percent of the world's population. Although the U.S. share of global economic production has dropped to about 24 percent, the United States achieves this production with just 4.2 percent of the world's population, including talented people from other countries who have come here to study, work, and otherwise contribute to the U.S. economy and society.

As discussed in Chapter 2, during the Cold War the United States could outspend its competitors to maintain its scientific, technological, commercial, and military advantages. The set of technologies that drove military competitiveness was relatively distinct from the set driving commercial products and markets. This separation between military and commercial technologies meant that foreign researchers could enjoy abundant opportunities to learn, create, and develop technologies in the United States without raising national security concerns. Under such circumstances, the dissemination of technologies of strategic importance to the United States could be clearly and vigorously controlled without substantially slowing the development of commercial technologies.

Finding 2: Notwithstanding changes in the competitive landscape, some technologies with specific national security value will always need to be protected from loss or unauthorized transfer. The challenge is to develop mechanisms for dealing with sensitive technologies originating from private-sector R&D activities that are later used for critical national security purposes.

Technologies with primarily national security applications that need to be protected include, most prominently, those developed by the U.S. government within a federally funded R&D center or other protected research environment. These technologies generally pose high barriers to entry, whether because of access to raw materials or requirements for specialized knowledge. Examples include nuclear weapons, stealth technologies, precision guidance, and other military applications of technology. As described more fully in Chapter 2, *applications* of such technologies as artificial intelligence (AI) and synthetic biology being used for military purposes also need to be protected, even if that is not the case for the basic technology itself. The government needs to be proactive in identifying which technologies fall into this category.

Finding 3: U.S. technology protection regimes are not sufficiently sensitive to the different, specific needs of open and restricted research environments.

Historically, leadership in scientific research and technology development has been built on the open exchange of information, broad participation, access to collaborators and research tools, and the flexibility to follow ideas to make discoveries. As noted above, however (see Finding 1), this level of openness was predicated on the assumption of U.S. economic and technological dominance. Restricted environments have the effect of limiting participation and access to information, reducing collaboration, and constraining innovation, all of which can unintentionally hamper progress. Limiting the adverse consequences of restrictions on R&D requires defining and maintaining different types of research environments that can match the restrictions being applied to the risks posed by a technology's dissemination.

The United States has many strengths that could support a revamped technology protection strategy, including the continued ability to attract the global talent required to maintain a disproportionate share of global GDP; the ability of companies and government to move quickly in developing, adopting, and integrating new technologies; the ability to leverage the strong national laboratory system, which includes both open and protected environments; and the ability to develop and exploit platforms that make use of resilient architectures to avoid the pitfalls of "perimeter defense" protection approaches. Important to keep in mind as well are the limitations imposed by the nation's values (e.g., openness), how other countries might exploit those values to their own advantage, and how the approaches they might take to this end can be mitigated. And if the United States is to maintain its technological leadership, it will have to ensure that it leverages its relative strengths, including both its robust private sector with open, competitive markets that enable businesses to use capital efficiently and the intellectual freedom of its universities.

Protecting U.S. strategic advantages goes beyond simply protecting the technology outputs of the innovation system. It is about risk management: striking the appropriate balance between protecting those outputs and promoting the conditions that favor discovery and innovation. The U.S. government has a vital role in this risk management effort. It defines which areas of technology (so called "critical technologies") are essential to U.S. interests; it defines the conditions for trust for participants in the development, production, or use of those critical technologies; and it accepts the risks of open research when those risks are outweighed by the benefits that accrue to the nation from remaining a leader in research, discovery, and the development of new technologies. The private sector performs similar risk assessments for technologies that are essential to commercial success, but the private sector cannot be expected to consider national security risks or the national interest.

> *Finding 4:* The United States now faces serious competition not just in the discovery and use of new science and new technologies but also in its capacity and capability to rapidly develop, adapt, and commercialize new technologies and, more broadly, take advantage of its overall innovation ecosystem.

> *Finding 5:* Foreign competition in science and technology is increasingly the product of other countries emulating the proven successful U.S. approach to R&D-based innovation, instead of being due to diversion or theft of U.S. technology.

As noted above and emphasized throughout this report, the United States today faces a competitive environment very different from that of the past. Other countries have been actively challenging the nation's long-standing leadership in fundamental research and technology innovation, most often by emulating the approach taken successfully by the United States: developing world-class R&D environments, attracting talent, and investing in and supporting technology development. Given the strong R&D ecosystems of other countries, an approach based on preventing competitors from developing many technologies similar to those developed in the United States by restricting access to or the use of those technologies is unlikely to succeed.

For such technologies as AI, synthetic biology, and advanced materials, the feedback between research and development is important to improving innovation and incorporating those technologies into products and processes. Many of these technologies also have low infrastructure costs relative to market size, lowering barriers for competitors. They are broad enablers of innovation and are incorporated into other technologies, functioning as platforms for many technological advances. Given declines in the costs of storage and transmission of data, AI is a notable example of a technology with limited infrastructure barriers. These factors increase the incentives for other countries to develop R&D systems that enable them to work on scientific and technological frontiers and enhance national competitiveness.

> *Finding 6:* The growth of systems-based technologies disrupts traditional approaches to technology protection. Because they are shared, such platforms cannot be protected using the historical approach of restricting use or knowledge without causing widespread problems with other technologies that share those platforms. Restricting use or knowledge can impact all phases of the technology life cycle, from development, to production, to use.

Risk management decisions have become more difficult as a result of the changes in technology and technology development described in Chapter 2. Technology products used to be largely discrete with well-defined purposes, but today's technologies are increasingly composed of systems that are highly shared, have multiple purposes and applications, and provide enormous advantages in scale and capability. The application of these highly shared technology platforms is often limited only by a developer's imagination, and control of such platforms runs the risk of inhibiting the ability to reap the benefits of leadership in these areas. The growing importance of platforms creates new policy problems for governing the development and use of the technologies that build on these platforms. Separation of technologies from the underlying platforms is often infeasible because of the loss of function or increase in costs that would result.

> *Finding 7:* The aggregation of myriad regulations and policies, implemented over decades and still enforced regardless of their continued effectiveness, has created a drag on the U.S. innovation engine. Since part of protecting advantage is ensuring the rapid creation of new technologies—especially in today's hypercompetitive global technology environment—this approach to technology protection has created self-inflicted barriers.

> *Finding 8:* The current approaches to risk identification and risk acceptance are diffuse, uncoordinated, and generally reactive rather than

proactive with respect to threats. Furthermore, protecting technologies through restrictions alone can have unintended consequences.

Because of the growth of strategic competition described in Chapter 3, as well as the importance to national defense of technologies driven primarily by considerations of global commercial application, the U.S. research community has seen a vast increase in the number and complexity of policies, processes, procedures, and requirements for the conduct of scientific and technological R&D. This increase, combined with the growing array of government stakeholders exercising authority, has created a set of overly complex—and sometimes conflicting—rules with major differences in requirements and their adoption and implementation across federal agencies. These rules limit the exchange of ideas, participation by other researchers, and international collaboration, slowing the pace of research and making research environments less attractive to talented people.

Finding 9: China has now fully emerged as the leading technological and economic competitor to the United States. China does not operate by the same international norms and standards that have guided the U.S. innovation ecosystem and its global engagements, and it has proven adept at exploiting weaknesses in the economic and governance structures and policies of the United States.

Finding 10: The historical approach to protecting technologies in the United States has generally consisted of unilateral (and in some cases, multilateral) reactions to external threats posed by adversaries. Risks in the new global R&D ecosystem cannot be managed effectively in this same manner without posing a new risk—that of inadvertently slowing the development and application of technologies and limiting competitive advantages.

Over the past two decades, China has systematically pursued strategies for dominating technology development in key areas. It has invested in R&D, sought to attract talent from other countries, and made massive investments in new technologies. China also does not play by the same rules as the United States. The Chinese government is deeply involved in commercial technology development; research outputs and data from competitors are subject to diversion or theft; foreign participation in the Chinese economy is limited and monitored; technology standards and regulations are managed to advantage domestic technologies; and markets are distorted to advantage domestic companies.

China's economy has grown substantially in the post–World War II period, surpassing that of Japan in 2010 to become the world's second-largest economy. Today, China's economy is estimated to represent more than 18 percent of the world's GDP, rivaling that of the United States (though China also has 18.5 percent of the world's population). China's R&D intensity has also ramped up in the last 5 years and, given a continuation of current trends, will surpass U.S. funding during the early part of this decade. China's rates of publication and patenting in science and engineering now exceed the rates among U.S.-based researchers. Also, as described in Chapter 4, China's ability to require companies to share information and its lack of strong privacy laws allow it to integrate and control information in ways that the United States does not, and its willingness to engage with authoritarian governments that the United States keeps at a distance provides it with access to markets and talent. China is willing to obtain technology through the acquisition of companies, foreign talent programs, and the theft of intellectual property, and it has learned that the United States will often react to such actions by establishing bureaucracies that will slow the U.S. innovation ecosystem.

> *Finding 11:* The strength of the U.S. research enterprise is dependent on access to sufficient numbers of high-level R&D scientists, engineers, and other technical personnel, both domestic and foreign. The United States is far from the point at which increases in the former displace the need for the latter.

> *Finding 12:* Disconnects and misalignments characterize the immigration policies the United States employs for international students and academic researchers and those it applies to technology workers.

> *Finding 13:* Foreign competitors are increasingly competing for international talent—both students and researchers. One way in which they are engaging in this competition is through incentives for the return of domestic workers trained in other countries.

Given the changing nature of today's technologies as described in Chapter 2, the ability to attract the best talent from around the world is essential in many cases to developing a technology and its applications. As a result, many nations have increased their efforts to develop, recruit, and retain talented people in their R&D ecosystems.

Compared with the U.S. immigration system, the systems of other English-speaking countries, such as the United Kingdom, Australia, and Canada, are more flexible and more focused on skills. High-skilled immigration is growing more rapidly in other Organisation for Economic Co-operation and Development (OECD) countries than in the United States, and the U.S. share of international students has been declining. Other countries have launched initiatives to recruit talent, including students who have been trained in the United States. While recent U.S. immigration reforms have allowed foreign students to work in this country longer (through so-called optional practical training opportunities) after completing their degrees, the number of temporary work visas is capped, even for those occupations in which workers are in high demand.

The strengths of U.S. graduate education are an important asset that the United States can use to its advantage. Talented people want to be around other talented people, creating a virtuous cycle of self-sustaining recruitment and retention and of continuous improvement. As described in Chapter 4, even though many STEM graduate students in the United States are from China and will eventually return home, the vast majority (more than 70 percent) of these students intend to stay in the United States and add to the innovative capacity of this country. As shown in Chapter 3, although the number of international graduate students in science and engineering enrolled at U.S. universities declined in 2020 because of the COVID-19 pandemic, this decline was smaller for science and engineering graduate students than for graduate students in other disciplines. Foreign-born workers remain an important part of the U.S. STEM workforce—making up 20 percent of that workforce overall and in some disciplines representing a much higher share.

6

Recommendations

Leadership in technological innovation advances the national security and economic interests of the United States. In an increasingly competitive and technology-dependent world, ensuring and protecting the nation's ability to lead in technological innovation is more important than ever before, and requires a new approach. The current approach is based on outmoded assumptions about the context in which technologies are developed and used. The first such assumption is that the United States enjoys an overwhelming advantage in the development of new technologies, and that this advantage can be protected by "outinnovating" adversaries and competitors. The second assumption is that strategically important technologies are discrete, with well-defined purposes. The third assumption is that these technologies continue to originate from federal laboratories and government-sponsored academic research and are subsequently commercialized for broader use. The final assumption is that the management of technology-related risks can be achieved primarily by protecting specific "critical technologies" from unauthorized use, possession, or production.

In today's extremely competitive global technology environment, these assumptions are no longer valid. In the committee's view, a fundamental shift in framing—one that goes beyond technology controls—is needed to protect U.S. technology advantages, setting the foundation for a new approach that has the following key objectives:

- *Maximization of strengths in science, research, and technology innovation.* The United States' greatest advantage over its competitors is rooted in an ability to be the first to develop and deploy new technologies, in cooperation with its allies, not in an ability to restrict access to technologies. Essential strategies for maximizing this advantage include promoting the scale and speed of the domestic research and technology innovation ecosystem; fostering a risk-taking environment to aid researchers and innovators; and attracting, retaining, and supporting the most talented science, engineering, and innovation workforce in the world. Recommendations 1 and 2 support these strategies.

- *Risk management based on threat identification and coordinated actions addressing the risks to U.S. technology leadership posed by these threats.* The current U.S. approach to managing the risks associated with technologies is based on restrictions on the possession, use, or manufacturing of these technologies, or restrictions on the knowledge or materials needed to develop them. Given the speed and scale of technology innovation and the growing trend of technologies originating in the private sector, the current risk management approach is of limited effectiveness and may in some cases be counterproductive. Instead, the U.S. government should focus on defining technology-related threats and vulnerabilities facing the United States, and then coordinate the implementation of effective strategies for responding to the resulting risks to U.S. technology leadership. Actions supporting these strategies might be taken in the public and private sectors; in multiple federal agencies; and if necessary, with international partners. Recommendation 3 addresses this issue.
- *A new multisector, multiorganizational, multinational approach to protection and assurance for the unique vulnerabilities associated with shared platforms.* Today's technology systems depend on, and are in many cases necessary components of, platforms for their functionality, production, or use. Platforms introduce new and shared vulnerabilities that can be exploited to misuse any technology on the platform. The codependency inherent in a shared platform means that restrictions or controls on the platform may disrupt everything using the platform—including beneficial uses that may enhance U.S. national security and competitiveness—creating very large-scale unintended consequences. The decentralized, and often international, governance systems that manage or control a platform may require coordinated federal action among multiple agencies responsible for standards, trade, international agreements, regulation, and law enforcement, or with private-sector entities or international partners. Recommendation 4 addresses this issue.

The following recommendations do not represent a comprehensive response to this need for a new framework for evaluating technology vulnerabilities. Instead, they represent important first steps toward a more effective approach to protecting the U.S. technological advantage based on such a framework.

MAXIMIZATION OF STRENGTHS IN SCIENCE, RESEARCH, AND TECHNOLOGY INNOVATION

The overall objective of the United States should be to maximize its strategic advantages in innovation and technology. Because scientific discovery and innovation favor broad and open participation, a key element of achieving this objective is to maximize the amount of work that can be performed appropriately in an open research environment, thereby promoting U.S. leadership in science and engineering, attracting top talent, and enhancing discoveries that lead to new technologies. For those specific cases in which an open environment is not suitable to protect U.S. interests, federal research and development (R&D) funders should make risk-informed decisions that clearly designate those specific cases requiring suitably restricted environments.

In conducting technology innovation, the United States benefits from having *both* open and restricted R&D environments. An open research environment is one with relatively few restrictions on participation, information sharing, or publication, but it does include basic requirements to ensure the integrity of the research process. Research, training, and teaching conducted in an open environment benefit the United States because they attract research talent, foster creative and innovative conditions for discovery, and speed the development of new ideas and technologies. Conducting this work in an open environment does pose a risk that knowledge, know-how, or results may flow to adversaries from the movement of either information or people. But for an innovation leader, the benefits of openness outweigh the risks for most R&D efforts because the risk of information loss is mitigated by the ability to innovate even newer technologies. An innovation leader can "run faster" than its competitors (IOM, NAS, and NAE, 1982).[1]

Most, though not all, research-related work is appropriate for an open environment. For certain specific uses, research, development, production, and related activities need to be confined to restricted environments that limit participation, collaboration, the sharing of information, and the dissemination of results to ensure that the knowledge, know-how, production, and use of a technology are limited to those entrusted to use the knowledge and information properly. In these cases, lowering the risk of disseminating a sensitive technology to adversaries outweighs the adverse impact of restrictions on the creativity and productivity of the work performed in such environments. U.S. policy should have the objective of striking the proper balance between these risks by designating the type of research environment most suitable for a given research activity.

It is the assessment of this committee that open research environments are not adequately defined in a way that protects the features most important to technology innovation and fundamental research. Generally, an open environment is one that is simply "not restricted"—for example, by classification or security clearance requirements. The key characteristics of an open research environment need to be defined specifically, in much the same way that a restricted research

---

[1] A 1982 National Academies report (commonly referred to as the Corson report) highlights that controls on scientific information "can be seen to weaken both military and economic capabilities by restricting the mutually beneficial interaction of scientific investigators, inhibiting the flow of research results into military and civilian technology, and lessening the capacity of universities to train advanced researchers" (IOM, NAS, and NAE, 1982, p. 3). The conclusion of the report is that open and free scientific communication is preferred because U.S. industry and military institutions will be able to "run faster" than U.S. adversaries (IOM, NAS, and NAE, 1982, p. 47).

environment is defined and protected. Thus defining these open research environments will clarify their importance and unique role in achieving the national objective of technological strength. Having this common definition will enable federal funding agencies to make an informed decision prior to an award as to whether the research they are funding is suitable for an open research environment or requires a restricted environment.

> *Recommendation 1:* The President, through an executive order, should clearly reaffirm that it is the policy of the United States that fundamental research, to the maximum extent possible, should remain unrestricted. In addition, the executive order should direct the Office of Science and Technology Policy, in coordination with federal agencies, to define criteria for open and restricted research environments within 120 days of issuance of the executive order. Furthermore, the executive order should direct federal agencies to designate the appropriate environment for work under a grant or contract prior to making the award, and to maximize the amount of sponsored work that can be performed in open research environments. In making this designation, agencies should state clearly that any restrictions or recommended restrictions apply only to the particular research grant or contract being funded, and not universally across the entire institution receiving the funding.

The proposed policy approach will put the federal government in the position of making an explicit risk acceptance decision on behalf of the nation. By designating certain research environments at universities or national laboratories as open, funding agencies will decide a priori what work can be performed in those environments despite the accompanying risks of disclosure. Similarly, research work that needs to be restricted, whether for commercial or national security reasons, will be explicitly designated for restricted environments, such as near-campus federally funded R&D centers, restricted government laboratories, commercial research facilities, or collaborative research centers between universities and companies. The committee notes that these policies are entirely consistent with those found in NSDD-189, which covers scientific, technical, and engineering information.[2]

---

[2] NSDD-189 states, "It is the policy of this Administration that, to the maximum extent possible, the products of fundamental research remain unrestricted. It is also the policy of this Administration that, where the national security requires control, the mechanism for control of information generated during federally-funded fundamental research in science, technology and engineering at colleges, universities and laboratories is classification. Each federal government agency is responsible for: a) determining whether classification is appropriate prior to the award of a research grant, contract, or cooperative agreement and, if so, controlling the research results through standard classification procedures; b) periodically reviewing all research grants, contracts, or cooperative agreements for potential classification. No restrictions may be placed upon the conduct or reporting of federally-funded fundamental research that has not received national security classification, except as provided in applicable U.S. Statutes" (White House, 1985, Section III).

In making determinations regarding protection, the burden of proof should be on the person or entity that wants to restrict, not on the group doing the work. The reason for restricting the work should be clearly stated, and the designation should be made as early as possible, ideally when proposals for funding are being solicited. The policy default should be that if no legitimate reason for restriction can be given, the work will be done in an open environment. This risk management approach is far preferable to generalized risk avoidance approaches, such as requiring background clearances for all researchers, "after-the-fact" restrictions applied when research has already been funded or performed, or pass-through restrictions from prime recipients to subcontractors.

DEVELOPING AND ATTRACTING TALENT

To compete effectively, the United States must lead in developing, attracting, and retaining top talent for research and innovation. World-class technology development and commercialization require the contributions of the best talent in the world. Since the end of World War II, the United States has enjoyed a comparative advantage over other countries by being the "go-to" destination for top foreign students and research talent. Talented scientists, engineers, and innovators are attracted to key features of America's open and democratic society: open and risk-embracing environments where they can pursue promising ideas; well-funded, world-leading academic and research institutions and technology companies; and the opportunity to be recognized or rewarded for their achievements. The readily available pool of international talent, however, has masked issues in training top domestic scientific and engineering talent, leaving the nation unprepared for efforts by other countries to reverse the flow of international talent to the United States.

The development of domestic talent will continue to be essential if the United States is to maintain its leadership in science and technology. As discussed in Chapter 3, the United States still lags other countries in preparing its citizens for participation in technology-intensive areas. Correcting this deficiency in domestic STEM (science, technology, engineering, and mathematics) education remains an urgent public policy objective if the United States is to continue to reap the benefits of being a leader in technology development.

At the same time, however, if the United States is to continue generating more than 20 percent of global gross domestic product (GDP) with only about 4 percent of the world's population, relying on talent from other countries will continue to be essential. The United States can no longer be complacent in assuming that it is the "default" choice for top global science and engineering talent. Other countries are aggressively competing for top students and STEM professionals, often by emulating the approaches that led to U.S. success in the past. Imposing excessive restrictions on foreign talent in research environments benefits U.S. competitors by dissuading talented people from coming to the United States, leading them to find other places to live and work. Therefore, in concert with efforts to expand domestic talent, the United States needs to aggressively expand the advantages it offers for international talent, including top academic research universities; open, well-funded, high-reputation, and innovative research environments; a venture-backed, entrepreneurial innovation system; and public–private R&D partnerships. China and other countries cannot match these advantages within the constraints of their systems.

Today, the United States faces growing competition for talent from other countries, including specific programs aimed at retaining or attracting away from the United States foreign or expatriate scientists or engineers. The current U.S. response to this competition is fragmented, defensive, and focused largely on restricting participation in foreign talent programs by U.S. citizens or institutions instead of encouraging recruitment of foreign talent. No coherent federal policy links efforts to strengthen domestic STEM education and training opportunities for U.S. citizens with efforts to attract top foreign talent as students or workers. The current policy approach does not adequately consider the need to strengthen or defend the features of the U.S. innovation system that have given it such long-standing advantages in attracting talent. While this committee was not asked to make recommendations in that area, it views this issue as so central to its charge that it advocates continued urgent attention to the issue at the highest levels of government.

Recommendation 2: The National Science Foundation (NSF) should fund and coordinate an effort to define those elements of the U.S. innovation system that are essential to developing, attracting, and retaining the top scientific, research, engineering, and innovation talent that is necessary for U.S. leadership in technology innovation. NSF should engage other federal science agencies, universities, research institutions, educators, and research-intensive companies in this effort. The agency should produce a report detailing its findings within 180 days of the start of the effort. Based on those findings, the Office of Science and Technology Policy should coordinate with federal research agencies, the Department of Homeland Security, and the Department of State to develop a national strategy for promoting leadership in science and technology through policies and programs aimed at developing domestic research talent, expanding opportunities for international research collaboration, and attracting and retaining top talent in the United States for training and employment.

Domestically, the same forces that attract students to finance, law, or medicine—a combination of personal fulfillment, contributing to the common good, a positive work environment, and a sustainable lifestyle—will attract students to STEM careers. Internationally, the United States needs to find new and better ways to encourage scientists, engineers, and their families to come to this country to work and live. Options include further aligning work visa levels with student visa levels and clearing pathways to citizenship for top international researchers.

IDENTIFICATION OF STRATEGIC TECHNOLOGIES AND COORDINATED RISK MANAGEMENT

U.S. policy should shift from an approach based on listing "critical" technologies, with associated restrictions, to one based on coordinated risk management. As explained in Chapter 2, the protection of technologies is complicated by the changing nature of modern technology. Technologies today are rarely isolated in a manner that allows for well-defined descriptions of a technology and its scope of use. Rather, technologies tend to be combined and interdependent, with emergent properties and uses as the component technologies evolve. The current technology landscape includes a more diverse set of users and uses of technology, including more multipurpose/multiuse technologies; a more diverse set of developers (including commercial and non-U.S. actors); and a more intertwined nature of technologies, markets, and applications. (Technology development and commercialization today are also different because of the rise of platforms, discussed in the next section.) The features of modern technology highlight the need to employ a comprehensive approach to managing the risks associated with strategically important classes of technology development or use.

A related consideration is how to respond to other countries, particularly those that may be geopolitical adversaries, when they identify specific technologies or technology areas for which they are seeking an advantage over the United States. Presently, the United States lacks a systematic policy approach to defining a suitable response strategy for this type of competition. In some cases, the strategies adopted by other countries, including China, pose risks to U.S. technology leadership that cannot be addressed effectively by the traditional approach of limited, laissez-faire government policy with regard to specific technology areas.

Historically, the U.S. approach to managing technology-related risks has focused on specifying "critical" technologies, generally based on their features or capabilities or on the consequences of their misuse, and then restricting access to those technologies or the means to produce them. Today, technology is ubiquitous, shared, and multipurpose; thus, the task of distinguishing technologies that pose a specific risk is very difficult and may result in identifying overly broad technology areas, further complicating efforts to manage risk because of unintended effects on innovation itself.

Alternatively, one could define a risk management approach that begins with identifying which actors using what means are attempting to use a particular technology against U.S. interests or technology leadership, and then defining strategies for addressing resultant risks. That approach requires expertise that goes beyond the nature of the technology to encompass the plans, actions, capabilities, and intentions of U.S. adversaries and other bad actors, thus involving experts from the intelligence, law enforcement, and national defense communities in addition to agency experts in the technology.

It is vital to U.S. economic and national security for the federal government to base its risk management strategy on specific, identified threats to

U.S. interests and leadership in technology innovation. Vital as well is that the actions resulting from any risk management strategy address the risks posed by these threats without having a deleterious effect on future scientific discovery, beneficial applications of a technology, and the nation's economic security.

Given the ubiquitous nature of today's technologies, the committee believes the process of identifying and coordinating federal strategies for managing technology-related risks must be coordinated across the federal government. The committee does not believe that this responsibility is presently well defined in any existing agency or at the interagency level within the White House. Just as many technologies are multidisciplinary in nature, government technology policy must be multiagency. As a starting point for action, the committee proposes using existing interagency mechanisms to begin an effort to define specific threat-informed considerations, to identify strategically important areas of technology-related risk, and to coordinate the development of strategies that can be deployed and coordinated by federal agencies to manage the identified risks to U.S. interests and technology leadership. Because the threats are to both economic and national security, the committee proposes a joint effort of the appropriate Cabinet-level councils within the White House. Given the scale of these technology and global competition issues, the committee believes it likely that an effective response may also require coordination of efforts with allies and other international partners. This approach would work best for a specific and limited number of strategically important technology areas, and would not replace other, routine forms of protection employed by government agencies or commercial enterprises.

> *Recommendation 3:* The National Security Council, the National Science and Technology Council, and the National Economic Council should develop and lead an interagency process for identifying and assessing threats or vulnerabilities of strategic significance to U.S. technology leadership and other national interests. For each threat, the process should include developing an associated risk management strategy and evaluation rubric for use by federal agencies in addressing the risk. The execution of these risk management strategies should be coordinated and overseen by the above interagency process to ensure a "whole-of-government" approach. The strategies resulting from this interagency risk management process should be
>
> - *proactive*, in that they define technology-related threats with national or economic security implications as early in the research and development process as possible;
> - *strategic*, in that they are based on global realities, including the plans, actions, intentions, and capabilities of adversaries, and on reasoned risk acceptance decisions about which technologies must, should, or cannot be protected;

- *timely*, in that they are based on current understanding of the associated threats and vulnerabilities and are adjusted as required;
- *integrated*, so that different mechanisms for technology protection, such as export controls, information classification, or decisions by the Committee on Foreign Investment in the United States, are directed and coordinated in such a manner as to effectively reduce or mitigate the risk;
- *adaptive*, with mechanisms for subjecting identified technology areas to regular reviews by integrated expertise in science, technology, and national security;
- *dynamic/repeating*, with a scheduled review to ensure that there have been no changes to the technology, the environment, or the actor(s) that would warrant a change in the threat status; and
- *assessed for adverse effects*, to ensure that they do not result in unnecessary and unintended barriers to U.S. innovation leadership.

The committee is aware that responsibility for risk management of technology-related threats belongs not solely to the federal government but also to the private sector and other actors in the national technology innovation system. Because the committee's charge was focused on actions for the federal government and for federally funded research, this report offers no specific recommendations on actions to enhance risk identification and management in the private sector. Nonetheless, the interagency process proposed in Recommendation 3 could be used to identify potential ways for the government to work with the private sector to improve risk management, including both collaborative efforts and consultation with industry before rules and regulations are passed.

Technologies are not static, and thus the means of protecting them cannot be static. A technology protection system must allow for continuing evaluation of the technology elements that need to be protected, with the mechanisms for managing overall risk being updated as necessary. While other risk management activities involving sensitive or critical technologies already occur at the agency level, it is important to define a whole-of-government risk management mechanism for those technologies that warrant it, either because of their nature or their potential applications, or because of the actions of U.S. adversaries. That mechanism should be managed in a manner that allows effective coordination across all affected agencies, including those involved with national security, law enforcement, trade, regulatory matters, international agreements, finance, science and technology, and standards setting.

TAILORED APPROACHES TO THE UNIQUE VULNERABILITIES RESULTING FROM SHARED PLATFORMS

The changing technological landscape has introduced new challenges to management of the risks posed by shared platforms and their supporting ecosystems. Current approaches to risk management assume that each technology is essentially independent of other technologies (regardless of whether they are in fact discrete or separable) and has a single purpose or small set of defined purposes. That assumption has been challenged by the emergence of what this report terms "platforms"—underlying technology systems that are foundational to the design, development, or use of other technologies. These platforms typically are highly shared, have multiple purposes and uses, and offer tremendous benefits through their scalability and adaptability. The most pressing omissions in current approaches to technology risk management in the United States involve these platforms.

Approaches to protecting diffuse multipurpose platforms differ from those for discrete, defined-purpose technologies. Sharing a common platform brings shared benefits as well as shared vulnerabilities and risks. At the country level, these shared vulnerabilities and risks affect the national interests of any country sharing the platform. Protecting these national interests typically requires governmental technology policies, such as government involvement in setting standards, regulations, or trade policies.

The committee does not believe that, at present, responsibility for identifying and managing the unique risks posed by these shared and powerful platforms is clearly established within the U.S. federal government, at either the federal agency level or the interagency level of the White House. Certain components of risk management suitable for application to platforms do exist in various agencies, but no agency has overall responsibility for coordination of these efforts. For example, the U.S. Department of Commerce contains separate bureaus and agencies for standards, protection of intellectual property, security of technology exports, telecommunications, and trade. These many efforts are subject to no coordination in accordance with a risk management strategy (i.e., one that identifies the risks, weighs those risks against the opportunities, and appropriately balances the two), either within the department or between the department and other federal agencies. This is but one example of the lack of ownership and cohesion that hampers U.S. efforts to engage with other global partners that share a platform and to address the shared vulnerabilities and risks.

The committee believes that the appropriate first steps in identifying strategically important platforms, defining the roles and responsibilities of federal agencies that pertain to those platforms, and developing coordinated risk management strategies covering their development, control, and use should be taken as part of a Cabinet-level interagency process. There are several reasons for this starting point: (1) at present, responsibilities in these areas are broadly spread across multiple departments; (2) input from the private sector and other key actors will be essential to gain a full understanding of how these shared platforms are managed, used, and changed; and (3) working with other countries and international bodies that share in these platforms will require the ability to engage at the international level.

Recommendation 4: The National Science and Technology Council, the National Security Council, and the National Economic Council should jointly develop a new policy framework for the identification of strategically important platforms and for the development of coordinated risk management strategies covering their development, control, and use. Elements of this new framework should include

- defining and designating specific technology platforms that are essential to U.S. interests;
- involving the private sector in specifying the technical features and requirements that should be included in platform development, such as performance standards for security, integrity, interoperability, control features, and user controls;
- developing a coherent, whole-of-government strategy for establishing and managing trust relationships among platform developers or users, including international governance mechanisms, use agreements, regulatory approaches, trade agreements, content requirements, and law enforcement cooperation agreements; and
- establishing a range of responses to security or trust problems related to the use of shared platforms, with participating agencies planning for and preparing appropriate "incident response" capabilities.

In today's interdependent, global innovation system, the greatest threat is that the United States will inadvertently weaken its innovation ecosystem while other countries continue to emulate the actions that have historically yielded U.S. advantages in technology development and commercialization. To counter this threat, the United States needs to protect and extend its ability to develop new technologies and apply those technologies to problems in both the military and commercial spheres. Protecting and strengthening this *ability* is vitally more important than protecting specific technologies.

References

Allen, G. C. 2019. *Understanding China's AI strategy: Clues to Chinese strategic thinking on artificial intelligence and national security.* Washington, DC: Center for a New American Security.

Aloe, J., and E. Guo. 2022, February 23. The US government is ending the China Initiative—Now what? *Technology Review.*

Alves Dias P., S. Amaroso, A. Annoni, J. M. Asensio Bermejo, M. Bellia, D. Blagoeva, G. De Parto, M. Doss, P. Fako, A. Fiorini, et al. 2019. *China—Challenges and prospects from an industrial and innovation powerhouse.* Luxembourg: Publications Office of the European Union.

Azoulay, P., B. Jones, J. D. Kim, and J. Miranda. 2022. Immigration and entrepreneurship in the United States. *American Economic Review: Insights* 4(1):71–88.

Brown, M. and P. Singh. 2018. *China's technology transfer strategy: How Chinese investments in emerging technology enable a strategic competitor to access the crown jewels of U.S. innovation.* Mountain View, CA: Defense Innovation Unit Experimental. https://nationalsecurity.gmu.edu/wp-content/uploads/2020/02/DIUX-China-Tech-Transfer-Study-Selected-Readings.pdf.

Buchholz, K. 2020, September 16. Where most students choose STEM degrees. *Statista.* https://www.statista.com/chart/22927/share-and-total-number-of-stem-graduates-by-country (accessed May 26, 2022).

Bush, V. 1945. *Science: The endless frontier.* Washington, DC: U.S. Government Printing Office. https://www.nsf.gov/od/lpa/nsf50/vbush1945.htm.

Capri, A. 2020. *Semiconductors at the heart of the US-China tech war: How a new era of techno-nationalism is shaking up semiconductor value chains.* Hong Kong: Hinrich Foundation.

Carlson, R. 2019, January 28. Presentation to the National Academies of Sciences, Engineering, and Medicine Committee on Safeguarding the Bioeconomy. Washington, DC.

CB Insights. 2022. *State of venture: Global, Q1 2022.* https://stats.oecd.org/Index.aspx?DataSetCode=MSTI_PUB.

Channa, G. 2021. How higher education became an important US export. *Issues in Science and Technology* 38(1):30–33.

Chia-Hsuan, Y., R. Nugent, and E. R. H. Fuchs. 2016. Gains from other's losses: Technology trajectories and the global division of firms. *Research Policy* 54(3):724–745.

Chik, H. 2021, July 16. China set to pass US on research and development spending by 2025. *South China Morning Post*. https://www.scmp.com/news/china/science/article/3141263/china-set-pass-us-research-and-development-spending-2025.

China State Council. 2017. *New generation artificial intelligence development plan*. Translated by New America. https://www.newamerica.org/cybersecurity-initiative/digichina/blog/full-translation-chinas-new-generation-artificial-intelligence-development-plan-2017.

CNCB-NGDC Members and Partners. 2020, November 11. Database resources of the National Genomics Data Center, China National Center for Bioinformation in 2021. *Nucleic Acids Research* 49(D1):D18–D28.

Cockburn, I. M., R. Henderson, and S. Stern. 2018. *The impact of artificial intelligence on innovation*. Working Paper 24449. Cambridge, MA: National Bureau of Economic Research.

Cohn, D., and N. Ruiz. 2017, July 6. *More than half of new green cards go to people already living in the U.S.* Washington, DC: Pew Research Center.

Commission on Scientific Communication and National Security. 2005. *Security controls on scientific information and the conduct of scientific research*. Washington, DC: Center for Strategic and International Studies.

Committee on New Models for U.S. Science and Technology Policy. 2020. *The perils of complacency: America at a tipping point in science and engineering*. Cambridge, MA: American Academy of Arts & Sciences.

CSET (Center for Security and Emerging Technology). 2021, May 12. *Translation of Outline of the People's Republic of China 14th Five-Year Plan for National Economic and Social Development and Long-Range Objectives for 2035*, translated by Etcetera Language Group, Inc.; edited by B. Murphy.

Cumbers, J. 2019, August 26. Trade deal or not, China is investing big in synthetic biology. *Forbes*.

Cunningham, M. A., and J. P. Geis II. 2020. A national strategy for synthetic biology. *Strategic Studies Quarterly* 14(3):49–80.

DHS (U.S. Department of Homeland Security). 2019. *LPR yearbook Tables 8 to 11 expanded*. https://www.dhs.gov/immigration-statistics/readingroom/lpr/table_8_to_11_expanded (accessed May 4, 2022).

Dobbs, R., J. Manyika, and J. Woetzel. 2015. *No ordinary disruption: The four global forces breaking all the trends*. New York: Public Affairs.

DoD (U.S. Department of Defense). 2022, February. *Securing defense-critical supply chains: An action plan developed in response to President Biden's Executive Order 14017*. https://media.defense.gov/2022/

Feb/24/2002944158/-1/-1/1/DOD-EO-14017-REPORT-SECURING-DEFENSE-CRITICAL-SUPPLY-CHAINS.PDF.

DOE (U.S. Department of Energy). 2020. *The state of the DOE National Laboratories, 2020 edition.* https://www.energy.gov/sites/default/files/2021/01/f82/ DOE%20National%20Labs%20Report%20FINAL.pdf.

El Karoui, M., M. Hoyos-Flight, and L. Fletcher. 2019. Future trends in synthetic biology—A report. *Frontiers in Bioengineering and Biotechnology* 7:175.

FBI (Federal Bureau of Investigation). 2019. *Executive summary: China: The risk to corporate America.* https://www.fbi.gov/file-repository/china-exec-summary-risk-to-corporate-america-2019.pdf/view.

Feldgoise, J., and R. Zwetsloot. 2020. *Estimating the number of Chinese STEM students in the United States.* Washington, DC: Center for Security and Emerging Technology.

Felin, T., and T. R. Zenger. 2014. Closed or open innovation? Problem solving and governance choice. *Research Policy* 43(5):914–925.

Fergusson, I. F., P. K. Kerr, and C. A. Casey. 2021. *The U.S. export control system and the Export Control Reform Act of 2018.* Washington, DC: Congressional Research Service.

Fialka, J. 2016, December 19. Why China is dominating the solar industry. *Scientific American.* https://www.scientificamerican.com/article/why-china-is-dominating-the-solar-industry (accessed May 26, 2022).

Finn, M. G., and L. A. Pennington. 2018. *Stay rates of foreign doctorate recipients from U.S. universities, 2013.* Oak Ridge, TN: Oak Ridge Institute for Science and Education.

Fuchs, E. H. R. 2014. Global manufacturing and the future of technology. *Science* 345(6196):519–520.

GAO (Government Accountability Office). 2022, April 19. *How artificial intelligence is transforming national security.* https://www.gao.gov/blog/how-artificial-intelligence-transforming-national-security (accessed May 27, 2022).

GenomeWeb. 2018, November 28. BGI research, China National GeneBank form synthetic biology alliance with Macquarie U. https://www.genomeweb.com/genetic-research/bgi-research-china-national-genebank-form-synthetic-biology-alliance-macquarie-u#.YsXZp-zMKi5.

Gil, Y., B. Selman, T. Dietterich, K. Forbus, M. desJardins, F. F. Li, K. McKeown, D. Weld, E. Bradley, A. Schwartz Drobnis, M. D. Hill, D. Lopresti, M. Matarić, and D. Parkes. 2019. *A 20-year community roadmap for artificial intelligence research in the US.* Washington, DC: Computing Community Consortium.

Gompert, D. C., and R. L. Kugler. 1996. *Rebuilding the team: How to get allies to do more in defense of common interests.* Santa Monica, CA: RAND Corporation. https://www.rand.org/pubs/issue_papers/IP154.html.

Goodrich, B. 2020a. *Research security revisited: COVID-19 & immigration.* Washington, DC: Consortium of Social Science Associations.

Goodrich, B. 2020b. *Foreign interference in the U.S. research enterprise and policy responses*. Washington, DC: Consortium of Social Science Associations.

Graham, A., K. Klyman, K. Barbesino, and H. Yen. 2021. *The great tech rivalry: China vs. the U.S.* Cambridge, MA: Belfer Center for Science and International Affairs.

Hass, R. 2020. *Stronger: Adapting America's China strategy in an age of competitive interdependence*. New Haven, CT: Yale University Press.

Hoadley, D. S., and K. M. Sayler. 2020. *Artificial intelligence and national security*. Washington, DC: Congressional Research Service.

Horowitz, M., P. Sharre, G. C. Allen, K. Frederick, A. Cho, and E. Saravalle. 2018. *Artificial intelligence and international security*. Washington, DC: Center for a New American Security.

Huggett, B. 2019. "Innovation" nation. *Nature Biotechnology* 37(11):1264–1276.

Huw, R., J. Cowls, J. Morley, M. Taddeo, V. Wang, and L. Floridi. 2021. The Chinese approach to artificial intelligence: An analysis of policy, ethics, and regulation. *AI & Society* 36:59–77.

Institute of International Education. 2021. *International students by academic level and place of origin, 2000/01-2020/21*. Open Doors Report on International Educational Exchange. http://www.opendoorsdata.org.

IOM, NAS, and NAE (Institute of Medicine, National Academy of Sciences, and National Academy of Engineering). 1982. *Scientific communication and national security*. Washington, DC: The National Academies Press. https://doi.org/10.17226/253.

Israel, E., and J. Batalova. 2021, January 14. *International students in the United States*. Migration Policy Institute. https://www.migrationpolicy.org/article/international-students-united-states-2020 (accessed June 12, 2022).

Jackson, J. K. 2020. *The Committee on Foreign Investment in the United States (CFIUS)*. Washington, DC: Congressional Research Service.

Kahn, S., G. La Mattina, and M. J. MacGarvie. 2017. "Misfits," "stars," and immigrant entrepreneurship. *Small Business Economics* 49(3):533–557.

Kania, E. B., and J. K. Costello. 2018. *Quantum hegemony? China's ambitions and the challenge to U.S. innovation leadership*. Washington, DC: Center for a New American Security.

Kania, E. B., and L. Laskai, 2021. *Myths and realities of China's military-civil fusion strategy*. Washington, DC: Center for a New American Security.

Khan, S. M. 2019. *Maintaining the AI chip competitive advantage of the United States and its allies*. Washington, DC: Center for Security and Emerging Technology.

Kim, D., and J. VerWey. 2019. *The potential impacts of the Made in China 2025 Roadmap on the integrated circuit industries in the U.S., EU and Japan*. Working Paper ID-061. Washington, DC: Office of Industries, U.S. International Trade Commission.

Kraemer, S. 2006. *Science and technology policy in the United States: Open systems in action*. New Brunswick, NJ: Rutgers University Press.

Kreps, S. 2021, November. *Democratizing harm: Artificial intelligence in the hands of nonstate actors*. Washington, DC: Brookings Institution. https://www.brookings.edu/research/democratizing-harm-artificial-intelligence-in-the-hands-of-non-state-actors (accessed May 27, 2022).

Kung, Y. T., Y. C. Du, W. T. Huang, C. C. Chen, and L. T. Ke. 1965. Total synthesis of crystalline bovine insulin. *Scientia Sinica* 14:1710–1716.

Kuo, S. 2020, March 5. Progress in importation of US equipment dispels doubts on SMIC's capacity expansion for mature nodes for now, says trend force. March 5, 2021 *TrendForce*. https://www.trendforce.com/presscenter/news/20210305-10693.html (accessed May 24, 2022).

Kwon, K. 2020, June 25. China reaches new milestone in space-based quantum communications. *Scientific American*. https://www.scientificamerican.com/article/china-reaches-new-milestone-in-space-based-quantum-communications (accessed August 30, 2022).

Langguth, J., K. Pogorelov, S. Brenner, P. Filkuková, and D. Thilo Schroeder. 2021, May 24. Don't trust your eyes: Image manipulation in the age of deepfakes. *Frontiers in Communication*. https://doi.org/10.3389/fcomm.2021.632317.

Lewis, J. A. 2021. *National security implications of leadership in autonomous vehicles*. Washington, DC: Center for Strategic and International Studies.

Li, J., H. Zhao, L. Zheng, and W. An. 2021. Advances in synthetic biology and biosafety governance. *Frontiers in Bioengineering and Biotechnology* 9:598087.

Liff, A., and G. J. Ikenberry. 2014. Racing toward tragedy? China's rise, military competition in the Asia Pacific, and the security dilemma. *International Security* 39(2):52–91.

Manyika, J., W. H. McRaven, A. Segal, A. Ackerson, D. Beck, N. F. Beim, J. Breyer, S. A. Denning, R. E. Dugan, R. Hoffman, A. Husain, N. Y. Lamb-Hale, E. S. Lander, M. Patel, D. J. Patil, L. R. Reif, E. Schmidt, R. M. Shah, L. D'Andrea Tyson, and J. Yang. 2019. *Innovation and national security: Keeping our edge*. Independent Task Force Report No. 77. New York: Council on Foreign Relations.

Mavroidis, P., and A. Sapir. 2021, April. China and the WTO: An uneasy relationship. *VOX EU*. https://voxeu.org/article/china-and-wto-uneasy-relationship.

McCarthy, N. 2017, February 6. The countries with the most STEM graduates. *Industry Week*. https://www.industryweek.com/talent/article/21998889/the-countries-with-the-most-stem-graduates (accessed May 26, 2022).

Mervis, J. 2022, February 28. Controversial U.S. China Initiative gets new name, tighter focus on industrial espionage. *Science*. https://www.science.org/content/article/controversial-u-s-china-initiative-gets-new-name-tighter-focus-industrial-espionage.

Moshasha, S. 2016. The rapid growth of synthetic biology in China. *SynBioBeta*. https://www.synbiobeta.com/read/the-rapid-growth-of-synthetic-biology-in-china.

Mowery, D. C., and N. Rosenberg. 1998. *Paths of innovation: Technological change in 20th-century America*. New York: Cambridge University Press.

NAE and NRC (National Academy of Engineering and National Research Council). 2013. *Positioning synthetic biology to meet the challenges of the 21st century: Summary report of a Six Academies Symposium Series*. Washington, DC: The National Academies Press. https://doi.org/10.17226/13316.

NAS, NAE, and IOM (National Academy of Sciences, National Academy of Engineering, and Institute of Medicine). 2007. *Rising above the gathering storm: Energizing and employing America for a brighter economic future*. Washington, DC: The National Academies Press. https://doi.org/10.17226/11463.

NAS, NAE, and IOM. 2009. *Ensuring the integrity, accessibility, and stewardship of research data in the digital age*. Washington, DC: The National Academies Press. https://doi.org/10.17226/12615.

NASEM. 2015. *Immigration policy and the search for skilled workers: Summary of a workshop*. Washington, DC: The National Academies Press. https://doi.org/10.17226/20145.

NASEM. 2016. *Optimizing the nation's investment in academic research: A new regulatory framework for the 21st century*. Washington, DC: The National Academies Press. https://doi.org/10.17226/21824.

NASEM. 2017. *Preparing for future products of biotechnology*. Washington, DC: The National Academies Press. https://doi.org/10.17226/24605.

NASEM. 2018a. *Open science by design: Realizing a vision for 21st century research*. Washington, DC: The National Academies Press. https://doi.org/10.17226/25116.

NASEM. 2018b. *Biodefense in the age of synthetic biology*. Washington, DC: The National Academies Press. https://doi.org/10.17226/24890.

NASEM. 2019a. *Adapting to the 21st century innovation environment: Proceedings of a workshop—In brief*. Washington, DC: The National Academies Press. https://doi.org/10.17226/25384.

NASEM. 2019b. *Quantum computing: Progress and prospects*. Washington, DC: The National Academies Press. https://doi.org/10.17226/25196.

NASEM. 2020a. *Review of the SBIR and STTR programs at the Department of Energy*. Washington, DC: The National Academies Press. https://doi.org/10.17226/25674.

NASEM. 2020b. *Safeguarding the bioeconomy*. Washington, DC: The National Academies Press. https://doi.org/10.17226/25525.

NASEM. 2021. *Advancing commercialization of digital products from federal laboratories*. Washington, DC: The National Academies Press. https://doi.org/10.17226/26006.

REFERENCES

NASEM. 2022. *Assessment of the SBIR and STTR programs at the National Institutes of Health.* Washington, DC: The National Academies Press. https://doi.org/10.17226/26376.

National Security Commission on Artificial Intelligence. 2021. *Final report.* Washington, DC: National Security Commission on Artificial Intelligence.

NCSES (National Center for Science and Engineering Statistics). n.d. Technical appendix: Publication output data and methodology. *Publications output: U.S. trends and international comparisons.* https://ncses.nsf.gov/pubs/nsb20206/technical-appendix (accessed August 25, 2022).

NCSES. 2022a, February 22. *National patterns of R&D resources: 2019–20 data update.* Data tables. National Science Foundation. https://ncses.nsf.gov/pubs/nsf22320.

NCSES. 2022b. *Survey of graduate students and postdoctorates in science and engineering.* Alexandria, VA: National Science Foundation. https://ncses.nsf.gov/pubs/nsf22319.

New American Economy. (2012, June 26). *Press release: New study reveals immigrants are behind more than three-quarters of patents from top ten patent-producing American universities.* https://www.newamericaneconomy.org/news/press-release-new-study-reveals-immigrants-behind-three-quarters-patents-top-ten-patent-producing-american-universities.

NFAP (National Foundation for American Policy). 2019. *Immigrants and Nobel Prizes: 1901-2019.* Arlington, VA: NFAP. https://nfap.com/wp-content/uploads/2019/10/Immigrants-and-Nobel-Prizes.NFAP-Policy-Brief.October-2019.pdf.

NRC (National Research Council). 2007. *Science and security in a post 9/11 world: A report based on regional discussions between the science and security communities.* Washington, DC: The National Academies Press. https://doi.org/10.17226/12013.

NRC. 2009. *Beyond "Fortress America": National security controls on science and technology in a globalized world.* Washington, DC: The National Academies Press. https://doi.org/10.17226/12567.

NRC. 2012. *Rising to the challenge: U.S. innovation policy for the global economy.* Washington, DC: The National Academies Press. https://doi.org/10.17226/13386.

NSB (National Science Board). 2019. Publications output: U.S. trends and international comparisons. *Science and Engineering Indicators.* Alexandria, VA: National Science Foundation. https://ncses.nsf.gov/pubs/nsb20214.

NSB. 2020a. Research and development: U.S. trends and international comparisons. *Science and Engineering Indicators 2020.* NSB-2020-3. Alexandria, VA: National Science Foundation.

NSB. 2020b. The state of U.S. science and engineering 2020. *Science and Engineering Indicators.* https://ncses.nsf.gov/pubs/nsb20201/global-science-and-technology-capabilities.

NSB. 2021. Publications output: U.S. trends and international comparisons. *Science and Engineering Indicators.* NSB-2020-6. Alexandria, VA: National Science Foundation. https://ncses.nsf.gov/pubs/nsb20206.

NSB. 2022a. The state of U.S. science and engineering 2022. *Science and Engineering Indicators.* NSB-2022-1. Alexandria, VA: National Science Foundation. https://ncses.nsf.gov/pubs/nsb20221.

NSB. 2022b. Higher education in science and engineering. *Science and Engineering Indicators 2022.* NSB-2022-3. Alexandria, VA: National Science Foundation. https://ncses.nsf.gov/pubs/nsb20223.

NSTC (National Science and Technology Council). 2016. *Advancing quantum information science: National challenges and opportunities.* Washington, DC: Executive Office of the President.

NSTC. 2022. *Guidance for implementing National Security Presidential Memorandum 33 (NSPM-33) on national security strategy for United States government-supported research and development.* Subcommittee on Research Security and Joint Committee on the Research Environment. Washington, DC: Executive Office of the President.

NVCA (National Venture Capital Association). 2022. *NVCA 2022 yearbook data pack: Public version.* https://nvca.org/wp-content/uploads/2022/03/NVCA-2022-Yearbook-PUBLIC-DATA-PACK.pdf.

OECD (Organisation of Economic Co-operation and Development). 2022. *Main science and technology indicators.* https://stats.oecd.org/Index.aspx?DataSetCode=MSTI_PUB (accessed May 31, 2022).

OMB (Office of Management and Budget). 1998, February 19. *Federal participation in the development and use of voluntary census standards and in conformity assessment activities.* Circular No. A-119. Washington, DC: OMB.

Open Doors. n.d. *International students: Enrollment trends.* https://opendoorsdata.org/data/international-students/enrollment-trends.

Parker, E., D. Gonzales, A. K. Kochhar, S. Litterer, K. O'Connor, J. Schmid, K. Scholl, R. Silberglitt, J. Chang, C, A. Eusebi, and S, W. Harold, 2022. *An assessment of the U.S. and Chinese industrial bases in quantum technology.* Santa Monica, CA: RAND Corporation. https://www.rand.org/pubs/research_reports/RRA869-1.html.

Pei, L., M. Schmidt, and W. Wei. 2011. Synthetic biology: An emerging research field in China. *Biotechnology Advances* 29(6):804–814.

People's Daily. 2017, May 2. China's biotech sector to exceed 4% of GDP by 2020: Authority.

Permanent Subcommittee on Investigations. 2019. *Threats to the U.S. research enterprise: China's talent recruitment plans.* Staff Report. Washington, DC: Committee on Homeland Security and Government Affairs, U.S. Senate.

Platzer, M. D., J. F. Sargent Jr., and K. M. Sutter. 2020, October 26. *Semiconductors: U.S. industry, global competition, and federal policy.* Washington, DC: Congressional Research Service.

President of the United States. 2020, June 4. Suspension of entry as nonimmigrants of certain students and researchers from the People's Republic of China. Proclamation 10043 as of May 29, 2020. *Federal Register* 85(108):34353–34355.

Ratnam, G. 2021, August 31. China's amassing of genomic data highlights global biotech race. *Roll Call.*

Research and Markets. 2021, March 17. *Semiconductor lithography equipment market—Growth, trends, COVID-19 impact, and forecasts (2021–2026).* https://www.reportlinker.com/p06036769/?utm_source=GNW (accessed May 26, 2022).

Riordan, M., and L. Hoddeson. 1997. *Crystal fire: The birth of the information age.* New York: Norton.

Roach, M., and J. Skrentny. 2019. Why foreign STEM PhDs are unlikely to work for US technology startups. *Proceedings of the National Academy of Sciences* 116(34):16805–16810. https://doi.org/10.1073/pnas.1820079 116.

Roach, M., H. Suermann, and J. Skrentny. 2019, November 24. Foreign-born entrepreneurial human capital in the US: The preference outcome gap. *VOX EU.* https://voxeu.org/article/foreign-born-entrepreneurial-human-capital-us.

Sargent, J. F., Jr. 2021. *Global research and development expenditures: Fact sheet.* Washington, DC: Congressional Research Service.

Saxenian, A. 2006. *The new argonauts: Regional advantage in a global economy.* Cambridge, MA: Harvard University Press.

Sayler, K. M. 2022a. *Emerging military technologies: Background and issues for Congress.* Washington, DC: Congressional Research Service.

Sayler, K. M. 2022b. *Defense primer: Quantum technology.* Washington, DC: Congressional Research Service.

Scharre, P., and R. Ainikki. 2020. *Defense technology strategy.* Washington, DC: Center for a New American Security.

Segal, S., and D. Gerstel. 2019. *Research collaboration in an era of strategic competition.* Washington, DC: Center for Strategic and International Studies.

Sganga, N. 2022, May 4. *Chinese hackers took trillions in intellectual property from about 30 multinational companies.* CBS News. https://www.cbsnews.com/news/chinese-hackers-took-trillions-in-intellectual-property-from-about-30-multinational-companies.

Shenzhen Institute of Advanced Technology. 2020, June 1. *Advanced Biofoundry Shenzhen.* http://english.siat.cas.cn/facility/202001/t20200106_229000.html.

Shenzhen Institute of Synthetic Biology. n.d. Introduction. *Shenzhen Biofoundry.* http://www.isynbio.org/institution_en.aspx.

Simcoe, T. 2012. Standard setting committees: Consensus governance for shared technology platforms. *American Economic Review* 102(1):305–336.

SIA (Semiconductor Industry Association). 2021, July. *Taking stock of China's semiconductor industry.* SIA Whitepaper. https://www.semiconductors.org/wp-content/uploads/2021/07/Taking-Stock-of-China%E2%80%99s-Semiconductor-Industry_final.pdf.

Snyder, T. D., ed. 1993. *120 years of American education: A statistical portrait.* Washington, DC: National Center for Education Statistics.

Sun, R., S. Gregor, and B. Keating. 2015. Information technology platforms: Definition and research directions. In *Proceedings of the 26th Australasian Conference on Information Systems*, edited by G. Deegan, F. Burstein, and H. Scheepers. Adelaide, University of South Australia. Pp. 1–17.

Tsang, A., and C. H. Poon. 2021, July 15. *China's 14th five-year plan: Research priorities and industrial policies.* HKTDC Research. https://research.hktdc.com/en/article/Nzk3NTY5NzUx.

USTR (Office of the U.S. Trade Representative). 2018, March 22. *Findings of the investigation into China's acts, policies, and practices related to technology transfer, intellectual property, and innovation under Section 301 of the Trade Act of 1974.* Washington, DC: Executive Office of the President.

Wetterstrand, K. A. 2021. *DNA sequencing costs: Data from the NHGRI Genome Sequencing Program (GSP).* www.genome.gov/sequencingcostsdata (accessed May 31, 2022).

White House. 1985. *National policy on the transfer of scientific, technical and engineering information.* National Security Decision Directive 189. https://irp.fas.org/offdocs/nsdd/nsdd-189.htm.

White House. 2021. *Building resilient supply chains, revitalizing American manufacturing, and fostering broad-based growth.* Washington, DC: The White House.

Wilson, J. R. 2016, October 1. How military harvests technology from commercial industry. *Military & Aerospace Electronics.* https://www.militaryaerospace.com/communications/article/16709009/how-military-harvests-technology-from-commercial-industry.

Zwetsloot, R., J. Feldgoise, and J. Dunham. 2020. *Trends in U.S. intention-to-stay rates of international Ph.D. graduates across nationality and STEM fields.* Washington, DC: Center for Security and Emerging Technology.

Zwetsloot, R., J. Corrigan, E. Weinstein, D. Peterson, D. Gehlhaus, and R. Fedasiuk. 2021. *China is fast outpacing U.S. STEM PhD growth.* Washington, DC: Center for Security and Emerging Technology.

APPENDIXES

Appendix A

Agendas

COMMITTEE MEETING 1
MARCH 4, 2021
VIA ZOOM

| Closed Session (12:00PM–2:00PM) |

| Open Session (2:00PM–4:00PM) |

2:00 PM **Welcome**
Susan Gordon, Duke University, and Former Principal Deputy Director of National Intelligence (Co-chair*)
Patrick Gallagher, University of Pittsburgh, and Former Director of the National Institute of Standards and Technology (NIST) (Co-chair*)

2:05 PM **Sponsor Perspectives**
Richard-Duane Chambers, Defense Advanced Research Projects Agency (DARPA)
Catherine Cotell, DARPA
Joshua Trapani, National Science Foundation (NSF)

3:00 PM **Overview of the Current Environment**
Richard Danzig, Center for a New American Security

4:00 PM **Break; Return to Closed Session**

| Closed Session (4:15PM–6:00PM) |

* Members of the committee identified with an asterisk (*).

COMMITTEE MEETING 2
APRIL 12, 2021
VIA ZOOM

Closed Session (1:00PM–4:00PM)

COMMITTEE MEETING 3: WORKSHOP ON SYNTHETIC BIOLOGY
MAY 13, 2021
VIA ZOOM

Open Session (1:00PM–6:00PM)

1:00 PM   **Welcome**
Susan Gordon, Duke University, and Former Principal Deputy Director of National Intelligence (Co-chair*)
Patrick Gallagher, University of Pittsburgh, and Former Director of NIST (Co-chair*)

1:05 PM   **Panel 1: Scientific and Policy History of Synthetic Biology**
Moderator: Richard Murray, California Institute of Technology*

Speakers: David Rejeski, Environmental Law Institute
David Walt, Harvard University

1:45 PM   **Panel 2: Current and Future Directions of Synthetic Biology: Viewpoints from Academia and Industry**

Part 1: Viewpoints from Academia
Moderator: Leroy Hood, University of Washington*

Speakers: Andrew Ellington, University of Texas at Austin
Pamela Silver, Harvard University

Part 2: Viewpoints from Industry
Moderator: Richard Murray, California Institute of Technology*

Speakers: Patrick Boyle, Ginkgo Bioworks
Steven Evans, BioMADE.org
Mostafa Ronaghi, Dynamics Special Purpose Corp.

APPENDIX A

3:45 PM	Break
4:00 PM	**Panel 3: Challenges and Promises for the Future** Moderator: Michael Imperiale, University of Michigan* Speakers: Drew Endy, Stanford University Richard Kitney, Imperial College London Diane DiEuliis, National Defense University
5:30 PM	**Final Thoughts/Additional Q&A**
6:00 PM	Adjourn

COMMITTEE MEETING 4: WORKSHOP ON MICROELECTRONICS
JUNE 10, 2021
VIA ZOOM

Open Session (1:00PM–5:30PM)

1:00 PM	**Welcome and Introductions** Susan Gordon, Duke University, and Former Principal Deputy Director of National Intelligence (Co-chair*) Patrick Gallagher, University of Pittsburgh, and Former Director of NIST (Co-chair*)
1:15 PM	**Panel 1: Historical and Current State of the Microelectronics Industry** Moderator: Robert Dynes, University of California, San Diego* Speakers: Kenneth Flamm, University of Texas at Austin Daniel Hutcheson, VLSI Research
2:15 PM	Break
2:30 PM	**Panel 2: The Implications of Offshore Production of Microelectronics and Slowing Moore's Law** Moderator: Gil Herrera, Sandia National Laboratories* Speakers: Darío Gil, IBM Lisa Porter, LogiQ T.J. Rodgers

4:00 PM	Break
4:15 PM	**Panel 3: Broad Recommendations about Federal Policies on Microelectronics** Moderator: Michael McQuade, Carnegie Mellon University* Speakers: William Chappell, Microsoft Britta Glennon, University of Pennsylvania John Manferdelli, Northeastern University
5:30 PM	Adjourn to Closed Session

Closed Session (5:30PM–6:00PM)

COMMITTEE MEETING 5: WORKSHOP ON ARTIFICIAL INTELLIGENCE
JULY 12, 2021
VIA ZOOM

Open Session (1:00PM–5:15PM)

1:00 PM	Welcome Susan Gordon, Duke University and Former Principal Deputy Director of National Intelligence (Co-chair*) Patrick Gallagher, University of Pittsburgh and Former NIST Director (Co-chair*)
1:05 PM	**Panel 1: Artificial Intelligence as a General Purpose Technology** Moderator: Richard Murray, California Institute of Technology* Speakers: Martial Hebert, Carnegie Mellon University Yolanda Gil, University of Southern California Daniela Rus, Massachusetts Institute of Technology
2:20 PM	Break

APPENDIX A

2:30 PM	**Panel 2: Threats Associated with Artificial Intelligence**
	Moderator: Michael McQuade, Carnegie Mellon University*
	Speakers: LtGen Michael Groen, Joint Artificial Intelligence Center
	Vidya Narayanan, Oxford University
	Stuart Russell, University of California, Berkeley
3:45 PM	**Break**
4:00 PM	**Panel 3: Recommendations for Artificial Intelligence as a Critical Technology**
	Moderator: Michael McQuade, Carnegie Mellon University*
	Speakers: Gilman Louie, Alsop Louie Partners
	Josephine Wolff, Tufts University
5:15 PM	**Adjourn to Closed Session**

Closed Session (5:15PM–6:00PM)

COMMITTEE MEETING 6
AUGUST 10, 2021
VIA ZOOM

Open Session (1:00PM–4:00PM)

1:00 PM	**Welcome**
	Susan Gordon, Duke University, and Former Principal Deputy Director of National Intelligence (Co-chair*)
1:15 PM	**Sponsor Perspectives: DARPA**
	Speaker: Carl McCants, DARPA
1:45 PM	**Break**
2:00 PM	**Critical Technologies and National Security**
	Speaker: Jason Matheny, Office of Science and Technology Policy

3:00 PM   **Government Control Mechanisms**
                 Moderator: Michael Imperiale, University of Michigan
                                Medical School**

                 Speaker: Gerald Epstein, National Defense University

4:00 PM   **Break; Adjourn to Closed Session**

Closed Session (4:15PM–6:00PM)

COMMITTEE MEETING 7
SEPTEMBER 13, 2021
VIA ZOOM

Open Session (1:00PM–4:00PM)

1:00 PM   **Welcome**
                 Susan Gordon, Duke University, and Former Principal Deputy
                      Director of National Intelligence (Co-chair*)
                 Patrick Gallagher, University of Pittsburgh, and Former
                      Director of NIST (Co-chair*)

1:05 PM   **Current State of Global Competition**
                 Speaker: John Culver

2:00 PM   **Assessing Risk in a New Global Environment**
                 Speaker: Donna Dodson

3:00 PM   **International Collaboration and Engagement**
                 Speakers: Farnam Jahanian, Carnegie Mellon
                                  University
                                Kent Fuchs, University of Florida

4:00 PM   **Break; Adjourn to Closed Session**

Closed Session (4:15PM–6:00PM)

COMMITTEE MEETING 8
OCTOBER 5, 2021
VIA ZOOM

Closed Session (1:00PM–6:00PM)

APPENDIX A

COMMITTEE MEETING 9
DECEMBER 13–14, 2021
VIA ZOOM

December 13, 2021
Closed Session (2:00PM–3:00PM)

December 14, 2021
Closed Session (4:00PM–6:00PM)

COMMITTEE MEETING 10
JANUARY 26, 2022
VIA ZOOM

Closed Session (12:00PM–6:00PM)

COMMITTEE MEETING 11
FEBRUARY 15, 2022
VIA ZOOM

Closed Session (12:00PM–1:00PM)

Appendix B

Biographies of Committee Members

PATRICK D. GALLAGHER (CO-CHAIR)

Patrick D. Gallagher is chancellor of the University of Pittsburgh, a role he has held since 2014 after having spent more than two decades in public service. In 2009, he was appointed to direct the National Institute of Standards and Technology. Dr. Gallagher has served on a number of National Academies committees, including the Committee on Scientific Assessment of Proposed U.S. Neutrino Experiments (2002), the Committee on Condensed Matter and Materials Research (2001–2005), the Committee on New Materials Synthesis and Crystal Growth (2007–2009), and the Government-University-Industry Research Roundtable (2009–2013). He is currently chair of the board of directors of the American Association of Universities (through October 2022). Dr. Gallagher received a Ph.D. in physics from the University of Pittsburgh in 1991.

SUSAN M. GORDON (CO-CHAIR)

The Honorable Susan M. Gordon served as principal deputy director of national intelligence from 2017 to 2019 and as deputy director of the National Geospatial-Intelligence Agency from 2015 to 2017. She joined the Central Intelligence Agency (CIA) in 1980 and served for 29 years, rising to senior executive positions in each of the agency's then four directorates: operations, analysis, science and technology, and support. In 1998, Ms. Gordon led the effort that culminated in the formation of In-Q-Tel, the CIA's venture arm. She is a fellow at Duke and Harvard universities; serves on several boards, including CACI International, Avantus Federal, BlackSky Technology, and MITRE; and advises several companies, including Microsoft Corporation. Ms. Gordon received a B.S. in zoology/biomechanics from Duke University in 1980.

ROBERT J. BIRGENEAU

Robert J. Birgeneau currently holds the Arnold and Barbara Silverman distinguished chair and faculty appointments in the departments of Physics, Materials Science and Engineering, and Public Policy at the University of California, Berkeley. He served as chancellor of the University of California, Berkeley, from 2004 to 2013. Dr. Birgeneau's research is primarily concerned with the phases and phase transition behavior of novel states of matter. Previously, he served as president of the University of Toronto and dean of science at the Massachusetts Institute of Technology. Dr. Birgeneau is a member of the National Academy of Sciences and a fellow of the Royal Society of London, the American Philosophical Society, and other scholarly societies. He has served as a member of the National Academies' Committee on Women in Science, Engineering, and Medicine and the Committee on Maximizing the Potential of Women in Academic Science and Engineering. Dr. Birgeneau received his Ph.D. in physics from Yale University.

ROBERT C. DYNES

Robert C. Dynes served as the 18th president of the University of California (UC) from 2003 to 2008. He is currently a professor of physics at UC San Diego, where he founded an interdisciplinary laboratory in which chemists, electrical engineers, and private industry researchers investigate the properties of metals, semiconductors, and superconductors. Dr. Dynes also served as the sixth chancellor of UC San Diego, following a 22-year career at AT&T Bell Laboratories, where he served as department head of semiconductor and material physics research and director of chemical physics research. He is a member of the National Academy of Sciences and currently serves as a member of the National Academies' Intelligence Community Studies Board. He chaired the Nuclear and Radiation Studies Board and the Committee on Disposal of Surplus Plutonium in the Waste Isolation Pilot Plant, among others. Dr. Dynes received his Ph.D. in physics from McMaster University in 1958.

DEBORAH FRINCKE

Deborah Frincke is associate laboratory director for national security at Sandia National Laboratories. Previously, she was associate laboratory director for national security sciences at Oak Ridge National Laboratory (ORNL), where she guided the research and development of science-based responses to complex threats that put public safety, national defense, energy infrastructure, and the economy at risk. Dr. Frincke joined ORNL from the National Security Agency (NSA), where she served in three roles between 2011 and 2020. As director of research at NSA from 2013 through early 2020, she led what is perhaps the largest in-house research organization in the U.S. Intelligence Community. In addition to being a founding member of the NSA board of directors, Dr. Frincke also served as the agency's science advisor and was the first NSA Innovation Champion. Prior to joining NSA, she served as full professor at University of Idaho and chief scientist for cybersecurity at Pacific Northwest National Laboratory; she also launched a successful cybersecurity startup company, TriGeo Network Systems. Dr. Frincke received a Ph.D. in computer science/security from the University of California, Davis, in 1992.

LEROY E. HOOD

Leroy E. Hood is senior vice president and chief science officer at Providence St. Joseph Health, and chief strategy officer, cofounder, and professor at the Institute for Systems Biology. Currently, he carries out studies in Alzheimer's disease, cancer, and wellness, and is pioneering a 1 million–patient genome/phenome project for Providence St. Joseph Health; he seeks to bring scientific (quantitative) wellness to the contemporary U.S. health care system. Dr. Hood is a member of the National Academy of Sciences, the National Academy of Engineering, and the National Academy of Medicine. He has served as a member on the National Academies' Division Committee for the Health and Medicine Division, the Committee on Human and Environmental Exposure Science in the 21st Century, and the Committee on Defining and Advancing the Conceptual Basis of Biological Science in the 21st Century, among others. Dr. Hood received his M.D. from the Johns Hopkins University School of Medicine in 1964 and his Ph.D. in biochemistry from the California Institute of Technology in 1968.

MICHAEL J. IMPERIALE

Michael J. Imperiale is Arthur F. Thurnau Professor in the Department of Microbiology and Immunology at the University of Michigan. With research focused on the study of DNA tumor viruses, he has made important contributions to our understanding of how these viruses regulate expression of their genes, how they contribute to oncogenesis, and how they interact with the infected cell in order to cause acute disease. Dr. Imperiale has served as a member of the National Academies' Committee on Science, Technology, and Law; chair of the Committee on Strategies for Identifying and Addressing Biodefense Vulnerabilities Posed by Synthetic Biology; and a member of the Committee on Ethical and Societal Implications of Advances in Militarily Significant Technologies That Are Rapidly Changing and Increasingly Globally Accessible. Dr. Imperiale received a Ph.D. in biological sciences from Columbia University in 1981.

J. MICHAEL MCQUADE

J. Michael McQuade serves as special advisor to the president at Carnegie Mellon University. Previously, he served as Carnegie Mellon's vice president for research. From 2006 to 2018, Dr. McQuade was senior vice president for science & technology at United Technologies Corporation, where he provided strategic oversight and guidance for research, engineering, and development activities that focused on a broad range of high-technology products and services for the global aerospace and building-systems industries. He has also held senior positions with technology development and business oversight at 3M, Imation, and Eastman Kodak, including vice president of 3M's Medical Division and president of Eastman Kodak's health imaging business. Dr. McQuade has broad experience managing basic technology development and the conversion of early-stage research into business growth. He served as a member of the President's Council of Advisors on Science and Technology, the Secretary of Energy Advisory Board, and the Defense Innovation Board, and is a member of the National Academies' National Science, Technology, and Security Roundtable. Dr. McQuade received a Ph.D. in physics from Carnegie Mellon University.

JUDITH A. MILLER

Judith A. Miller is an independent consultant. She served as general counsel to the Department of Defense from 1994 to 1999. Ms. Miller was founder and co-chair of, and is now senior advisor to, the ABA Cybersecurity Legal Task Force; she was a member of the Defense Science Board's Task Force on Cyber Deterrence, and she has also co-chaired the Defense Science Board's Task Force on Department of Defense Dependencies on Critical Infrastructure. Ms. Miller was senior vice president, general counsel, and member of the board of directors of the Bechtel Group from 2006 to 2010. Prior to joining the Bechtel Group, she was a partner with Williams & Connolly LLP in Washington, DC. Currently, Ms. Miller serves as a member of the National Academies' Committee on Science, Technology, and Law. She previously chaired the Committee on Section 230 Protections: Can Legal Revisions or Novel Technologies Limit Online Misinformation and Abuse?, and served as a member of the Committee on Scientific Communication and National Security; the Committee on Science, Security, and Prosperity in a Changing World; and the Roundtable on Scientific Communication and National Security. Ms. Miller earned a J.D. from Yale Law School in 1975.

RICHARD M. MURRAY

Richard M. Murray is Thomas E. and Doris Everhart professor of control & dynamical systems and bioengineering at the California Institute of Technology. His group researches the application of feedback and control to networked systems, with applications in biology and autonomy. His current projects include novel control-system architectures, biomolecular feedback systems, and networked control systems. Dr. Murray is a member of the National Academy of Engineering (NAE) and serves on the NAE Awards Committee.

Previously, he served as chair of the National Academies' Aerospace Engineering Search Committee, member of the Committee on Defense Research at Historically Black Colleges and Universities and Other Minority Institutions, chair of the Committee on Future Biotechnology Products and Opportunities to Enhance Capabilities of the Biotechnology Regulatory System, and co-chair of the Forum on Synthetic Biology. Dr. Murray received a Ph.D. in electrical engineering and computer sciences from University of California, Berkeley, in 1990.